Design, Development and Fabrication of Solar powered High Altitude Long Endurance (HALE) Unmanned Flying Vehicle (UAV)

By

Ahmad Bilal

Shuja Ur Rehman

ABSTRACT

Design, Development & Fabrication of Solar Powered High Altitude Long Endurance (HALE) Unmanned Aerial Vehicle (UAV)

The objective of this project is to design a High Altitude and Long Endurance (HALE) UAV which could be used for military as well as commercial purposes. The aircraft is required to climb at high altitude and to stay in air for a long time. To achieve the mentioned objectives, a light and low drag flying wing configuration was chosen with solar cells as major power source.

The project includes the conceptual and detail designing of the aircraft with stability, propulsion, structural and fluid analysis, and the fabrication of a prototype flying model for the demonstration of the project. Some iterative processes and optimizations were used to converge onto the most efficient design. A mechanism for continuous flight was developed using solar cells to achieve the required endurance.

The project has various applications that include Remote sensing, Reconnaissance, Communication and Atmospheric studies. In a nutshell, the project can be a substitute for a satellite and is very favorable economically as compared to a satellite.

Acknowledgments

First of all we would like to thank Almighty Allah for His blessings on us to make us capable enough to complete the project in time. Further, our project titled with *"Design, Development and Fabrication of Solar Powered High Altitude Long Endurance (HALE) Unmanned Aerial Vehicle (UAV)"* is the result of efforts of many people specially our project advisor **Mr. Sher Afgan**, whose theoretical and practical knowledge, experience, interest, guidance, encouragement and most importantly the motivation towards our project has given us all to complete the project.

We sincerely want to express our deep gratitude towards our Vice Chancellor Brig.Imran Rehman who always encouraged us to do this project. Then our Head of Department Dr.Naeem whose administrative cooperation also played an important role in the completion of our project. Further, we want to acknowledge Mr.Salman Nazeer whose guidance and cooperation in fabrication of prototype model also worth a lot in this design project.

We are also thankful to the Department of Aeronautics & Astronautics, Institute of Space Technology, in which coordination of so many people enabled us to finally complete the final year design project specifically Mr.Behram Khan, Mr.Ali Asif and Mr.Naveed.

Last but not least, our colleagues whose advices, appreciation and support also played an important role in our project specially Ms.Zainab Naseer, Mr.Mustafa Ali Ghazi, Mr.Qasim Raza, Mr.Faraan and Mr.Zubair Sajid.

SAIST Team

Table of Contents

List of Figures

List of Tables

1.0 Introduction

High Altitude Long Endurance (HALE) air vehicles have been the focus of significant research and development efforts for decades. The achievement of a solar powered aircraft capable of continuous flight was still a dream some years back, but this great challenge has become feasible today. In fact, significant progresses have been realized recently in the domains of flexible solar cells, high energy density batteries, miniaturized MEMS and CMOS sensors, and powerful processors.

The concept is quite simple; equipped with solar cells covering its wing, it retrieves energy from the sun in order to supply power to the propulsion system and the control electronics, and charge the battery with the surplus of energy. During the night, the only energy available comes from the battery, which discharges slowly until the next morning when a new cycle starts.

The state of the art has been advanced to enable higher operational altitudes, longer durations with greater payloads, and increased autonomy. Applications for these vehicles include atmospheric studies, remote sensing, surveillance, and local communication coverage. The scope of electric powered Unmanned Aerial Vehicle (UAV) missions is primarily constrained by the relatively low flight endurance that is characteristic of electric powered UAVs. The purpose in utilizing solar energy to power UAVs is to increase UAV endurance capabilities and ultimately achieve indefinite sustained flight.

So, we the SAIST design team has worked on designing of solar powered HALE UAV capable of 12hrs endurance and having an absolute ceiling of 80,000ft.A scale down prototype is also a part of the project in order to brief the concept of HALE UAV.

2.0 Problem Statement

It is proposed to **design, develop and fabricate a solar powered high altitude long endurance (HALE) unmanned flying vehicle**. The flying vehicle should be capable of taking off from the ground and reach a high altitude. As the flying vehicle is intended to provide surveillance or to provide local communication coverage for a particular area, it's desired to have endurance of about 12 hours and cruise speed of 74ft/s at 60,000ft. The weight constraint of the flying vehicle is 300lbs which also includes the payload weight of 50lbs. The propulsion system will include efficient electric powered motors, energy storage unit (batteries), a solar array and a power management unit.

The flying vehicle would be so designed to get the maximum solar input from the sun during the day time. For that purpose, the wing area of the vehicle would be optimized to largest as possible which will give a smaller wing loading. The power management unit would basically be a controller which would manage the amount of potential difference produced by the solar array by distributing it such that a part of the energy produced runs the motors, part of it charges the batteries for night flight and the rest used to run the avionics.

The mission profile of the vehicle includes take off, climb, cruise, loiter and landing as its major phases. According to the design requirements, the loitering phase would be the largest phase of flight of the flying vehicle and in this phase, the major concern of design would be the level flight because whether it is desired to achieve surveillance at low or perform as a high altitude communication platform, a solar aircraft capable of flying continuously needs to fly at constant altitude.

Chapter 3 | Executive Summary

3.0 Executive Summary

This project report describes the overall design methodology, analysis and validation of design as well as the manufacturing of Solar powered HALE UAV. Design constraints and project requirements have been optimized as to converge on the most efficient performance, aerodynamics and stability. The project has been done in different phases that mainly includes the conceptual design phase, analysis phase and detailed design phase that further includes the fabrication of scaled down prototype flying model.

Conceptual design involves the selection of best configuration with respect to the given design requirements and then initial sizing that is verified by analysis of wing loading variation in each segment of mission profile and then final aircraft sizing has been done. A low drag flying wing configuration has been selected and thus a number of figure of merits have been established for airfoil selection and comparison of different flying wings airfoils finally converged towards a reflex cambered airfoil.

Analysis phase involves the aerodynamic calculations and aerodynamic parameters variation curves. Gridgen and FLUENT software packages have been used for grid generations and flow simulation and their results are compared with that of manual calculations. Performance analysis involves the optimization of design variables in order to converge on the best aircraft flight conditions fulfilling the design requirements. Stability analysis has been done to insure the stable flight profile during all segments of mission profile. Structural analysis has also been done that includes the prediction of overall loading conditions, structural considerations and the complete structural analysis using ANSYS software package the results of which has been used in structural design process.

Propulsion involves the selection of propulsive components using optimization algorithm along with power management from the solar panels. Power management involves the detailed calculation of power accumulation from the solar panels and distribution between all propulsive components. The efficiencies have also been evaluated and required power has been managed by using optimized model of propulsion mechanism.

Detailed design phase involves a comprehensive design consideration of structural components, manufacturing processes, material selection and fabrication facility. A static model for demonstration of HALE UAV configuration has also been developed and presented.

A scaled down prototype flying model has been developed for concept demonstration of solar powered high altitude long endurance unmanned flying vehicle.

Chapter 4 | Project Management

4.0 Project Management

Project Management is the most important part in the completion and execution of the complete project. Project has been managed by dividing it into three phases and further making task distribution between group members of the SAIST design team. Time line has also been planned in accordance with many considerations and constraints.

4.1 Project Phases

The project has been divided into three phases as

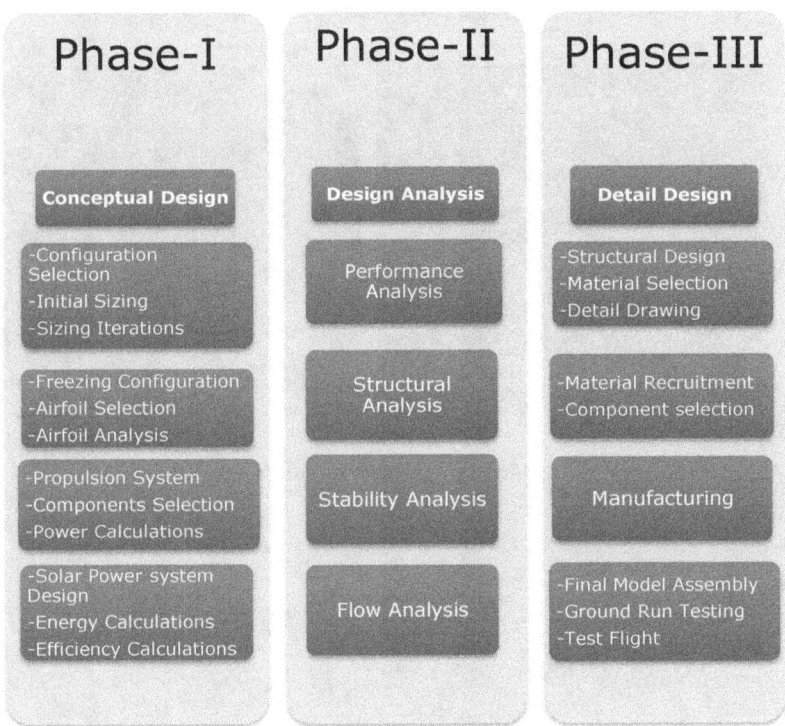

[Fig 4.1: SAIST Design Project Phases]

4.1.1 Project Management Team

SAIST project has been managed between the team members and within the allocated time. The effort in project management is to complete the project by proper planned meetings with the advisor and by proper task distribution between project members.

The SAIST HALE UAV design team and task distribution is as

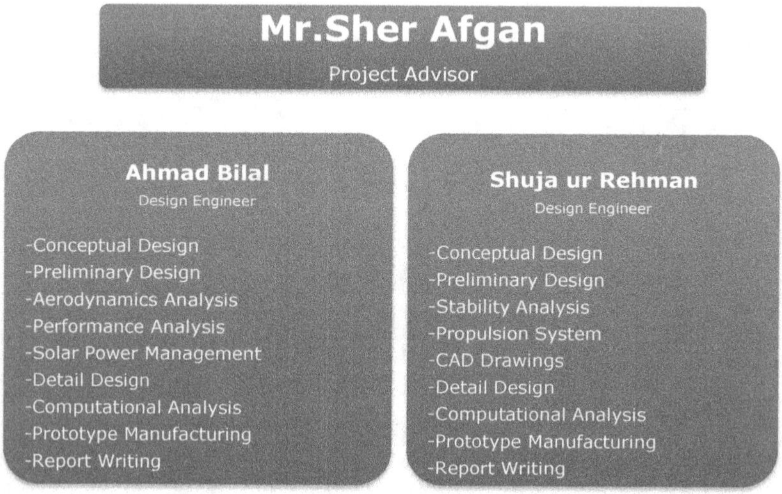

[Fig 4.2: SAIST Project Team & Task Management]

4.1.2 Gantt Chart

The Gantt chart shown below is basically describing the timeline of the complete project.

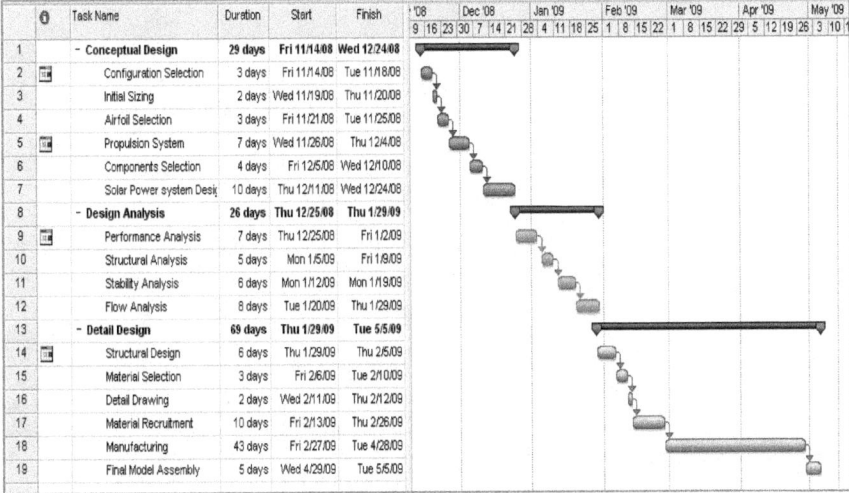

	❸	Task Name	Duration	Start	Finish
1		− Conceptual Design	29 days	Fri 11/14/08	Wed 12/24/08
2	▦	Configuration Selection	3 days	Fri 11/14/08	Tue 11/18/08
3		Initial Sizing	2 days	Wed 11/19/08	Thu 11/20/08
4		Airfoil Selection	3 days	Fri 11/21/08	Tue 11/25/08
5	▦	Propulsion System	7 days	Wed 11/26/08	Thu 12/4/08
6		Components Selection	4 days	Fri 12/5/08	Wed 12/10/08
7		Solar Power system Desig	10 days	Thu 12/11/08	Wed 12/24/08
8		− Design Analysis	26 days	Thu 12/25/08	Thu 1/29/09
9	▦	Performance Analysis	7 days	Thu 12/25/08	Fri 1/2/09
10		Structural Analysis	5 days	Mon 1/5/09	Fri 1/9/09
11		Stability Analysis	6 days	Mon 1/12/09	Mon 1/19/09
12		Flow Analysis	8 days	Tue 1/20/09	Thu 1/29/09
13		− Detail Design	69 days	Thu 1/29/09	Tue 5/5/09
14	▦	Structural Design	6 days	Thu 1/29/09	Thu 2/5/09
15		Material Selection	3 days	Fri 2/6/09	Tue 2/10/09
16		Detail Drawing	2 days	Wed 2/11/09	Thu 2/12/09
17		Material Recruitment	10 days	Fri 2/13/09	Thu 2/26/09
18		Manufacturing	43 days	Fri 2/27/09	Tue 4/28/09
19		Final Model Assembly	5 days	Wed 4/29/09	Tue 5/5/09

[Fig 4.3: Gantt chart]

Chapter 5 | SAIST Design Procedure

5.0 SAIST Design Procedure

The sequence of processes followed in SAIST design is as

Conceptual configuration selection of solar powered High altitude long endurance unmanned flying vehicle

Historical analysis and case studies of solar powered/HALE unmanned flying vehicles

Establishment of requirements for our particular aircraft to accomplish specific mission

Conceptual comparison of different configurations according to established requirements

Selection of baseline configuration

Preliminary design, Aerodynamic, Performance, structure and stability Analysis

Solar Power Management, Detail Design, Structural Consideration, Material Selection

Computational Analysis (Fluid, Structural), CAD Drawings

Prototype Manufacturing, Material & Component Recruitment

Ground Testing Of Prototype Model

Pre-Flight Checks

Test Flight, Modification after Test Flight

Final flight demonstration

[Fig 5.1: SAIST Design Procedure]

5.1 Previous Research on Solar powered vehicles

Before starting with the conceptual design phase an overview of previous solar powered aircrafts was made for conceptual evaluation of SAIST HALE UAV.

In the selection of configuration the most important factor is that of the historical trend or historically validated configurations in solar powered HALE flying vehicles. Initially a case study of different solar powered vehicle was made which is as follows:

Sunrise

Sunrise I - the first ever solar powered drone took to the sky at Camp Irwin Bicycle Lake, California on November 4, 1974. It was built by the U.S. Company Astro Flight. The Sunrise I used over 1000 solar cells located on the wings to produce about 450 watts of power. Sunrise I had a 32 feet wing span and weighed 26 pounds and had

a service ceiling of about 20,000 feet on a clear day. A year later an improved version named **Sunrise II** took its first flight. The Sunrise II with its 4480 solar cells delivered over 600 watts and weighed only 4 lbs (1.8kg). With more power and a lighter weight the climb rate of the Sunrise II reached about 300 ft per minute or 20,000 ft per hour and the service ceiling was estimated to be about 75,000 feet (various command and control issues limited this number considerably).

[Fig 5.1: Sunrise]

Solar Challenger

The next step in solar aviation came in 1980. The U.S. Company AeroVironment attempted to build a solar-powered piloted aircraft that could fly from Paris, France to England. The first prototype, called the **"Gossamer Penguin",** was fragile and not very airworthy, but led the way to an improved version called the **"Solar Challenger"**.

[Fig 5.2: Gossamer]

[Fig 5.3: Solar Challenger]

The Solar Challenger had a wingspan of 14.3 meters (47 feet) and a weight of 200lbs (90kg). It had 16,128 photovoltaic cells that covered the wings, with a total output power of a whooping 2,600 watts. The Solar

Challenger was capable of reaching an altitude of 12,000 ft, and in July 1981 the aircraft became the first to cross the 163 miles (262km) distance from Paris to Manston in the UK only using solar power.

Pathfinder

The **Pathfinder** reached an altitude of 50,500 feet on September 11, 1995, setting a

new altitude record for solar-powered aircraft. Three years later modifications to the Pathfinder resulted in a longer-winged version called **Pathfinder-Plus**. On August 6, 1998, the modified aircraft was flown to a record altitude for propeller-driven aircraft of 80,201 feet. The goal of the flight was to validate new solar,

[Fig 5.4: NASA's Pathfinder]

aerodynamic, propulsion and systems technology developed for the Pathfinder's successor, the Centurion, which was designed to reach and sustain altitudes within the 100,000-foot range. The Pathfinder-Plus incorporated several improvements to the original Pathfinder. One of the more noticeable features being stronger solar cells with 19% efficiency (the original Pathfinder had only 14%) developed by Sun Power Corp. in Sunnyvale, California. The new silicon solar cells boosted the maximum potential power from about 7,500 Watts on Pathfinder to about 12,500 Watts on Pathfinder-Plus.

Centurion

The Centurion, like its immediate predecessors Pathfinder and Pathfinder-Plus, was a lightweight, solar-powered, remotely piloted flying wing that demonstrated the technology of applying solar power for long-duration, high-altitude flight. The Centurion had more than twice the wingspan of the Pathfinder and about 2/3 more than the Pathfinder-Plus (98 feet in the Pathfinder and 206 feet Centurion). The Centurion's test flights were done by NASA in 1998 and a year later was modified and renamed the Helios.

[Fig 5.5: Centurion]

Helios

Helios is the most ambitious of NASA's solar UAV projects to date. With a 247-foot wingspan (greater than a 747 Jumbo Jet), 62,120 bi-facial solar cells and a projected maximum flying altitude of 100,000 feet, the Helios is the peak of two and a half decades of solar aviation research. Like its predecessors the Helios was built by AeroVironment and has been used on the Environmental Research Aircraft and Sensor

[Fig 5.6: NASA's Helios]

Technology (ERAST) program. The Helios applies hybrid technology – solar energy using photovoltaic cells by day and fuel cells by night. The Helios was designed to be the forerunner of high-altitude unmanned aerial vehicles that could fly ultra-long duration environmental science or telecommunications relay missions lasting for weeks or months without using consumable fuels or emitting airborne pollutants. On August 13, 2001, Helios demonstrated its capability when it reached an unofficial altitude record for non-rocket-powered aircraft of 96,863 feet. Unfortunately, on June 26, 2003 during a test flight over the Pacific Ocean near Kauai, Hawaii the Helios prototype was lost due to a structural failure caused by control problems.

Helipat

From 2000-2003 a team from the Politecnico de Torino in Italy together with a team from the University of York in the U.K developed a concept for the Heliplat - a Very-Long Endurance Solar Powered Autonomous Aircraft (VESPAA). Heliplat and other VESPAA UAVs could play the role of a "pseudo satellite", with the advantages of being closer to the ground, more flexible and at a cost much less than a real satellite.

[Fig 5.7: Helipat]

Heliplat-like HALE flying above a major city will be able to cover an area 1000 km across, and process a predicted 425,000 cell phone conversations simultaneously. This means a user community of 8.5 million per unit (although this does not take into account data transmission).

SOLITAIR

While the Heliplat was designed to have a large wingspan (70 meters), the German Aerospace Center (DLR) is currently designing a much smaller HALE known as the **SOLITAIR**. The SOLITAIR will only have a 5.2 meter wingspan and four solar panel sections embedded into its body that will use small engines to rotate towards the sun. Like the Heliplat, the SOLITAIR is also being developed in order to replace some of the tasks currently done by satellites.

[Fig 5.7: SOLITAIR]

Sky-Sailor

Sky-Sailor was in fact the first step towards designing a feasible solar powered aircraft to study Mars under collaboration with European Space Agency. It is equipped with 216 silicone solar cells delivering up to 90 W of energy having a wing span of 3.2 meters and weights up to 2.6kg.

[Fig 5.8: Sky- Sailor]

Zephyr

Zephyr, along with solar power, uses 'low drag aerodynamics' to fly for months at an altitude of 132,000 feet. It's yet another high altitude communication platform with 12-meter solar cells equipped wingspan churning out 1 kW of power to five motors that drag it to 70 meters per second (155 mph).Zephyr was manufactured by QinetiQ having a span of 12 meters and weighs up to 14 kg .It can fly at 70 meters per second.

[Fig 5.9: Zephyr]

SoLong

The **SoLong** is an electric-powered UAV (unmanned aerial vehicle) that collects solar energy from photo-voltaic arrays laminated into its wings. It uses energy so efficiently that it can fly all night on energy it gathers from the sun during the day. Remaining aloft for two nights is the milestone for sustainable flight. One night is possible just by discharging

[Fig 5.10: SoLong]

the batteries, but two or more nights means that the plane has to fully recoup and store the energy used at night while flying in the sunlight the following day. Once that is achieved, the cycle can repeat continually, and keep the plane airborne indefinitely.

Solar Powered bird

The solar-powered bird is literally an eagle and apparently is world's first flapping-wing unmanned aircraft that runs on solar energy. Designed by researchers at the University of Missouri-Rolla with the help of NASA, the aircraft uses thin-film solar arrays and special shape changing material hence making it more flexible and efficient than the Mars Rovers.

It can regain its altitude, up to 30,000 to 40,000 feet, by simply flapping its wings and is a perfect flying machine to inspect outer space.

Venus

A **Venus** exploration aircraft, sized to fit in a small aero shell for a "Discovery" class scientific mission, has been designed and analyzed at the NASA Glenn Research Center. For an exploratory aircraft to remain continually illuminated by sunlight, it would have to be capable of sustained flight at or above the wind speed, about 95 m/sec at the cloud-top level. The analysis concluded that, at typical flight altitudes above the cloud layer (65 to 75

[Fig 5.11: Venus]

km above the surface); a small aircraft powered by solar energy could fly continuously in the atmosphere of Venus.

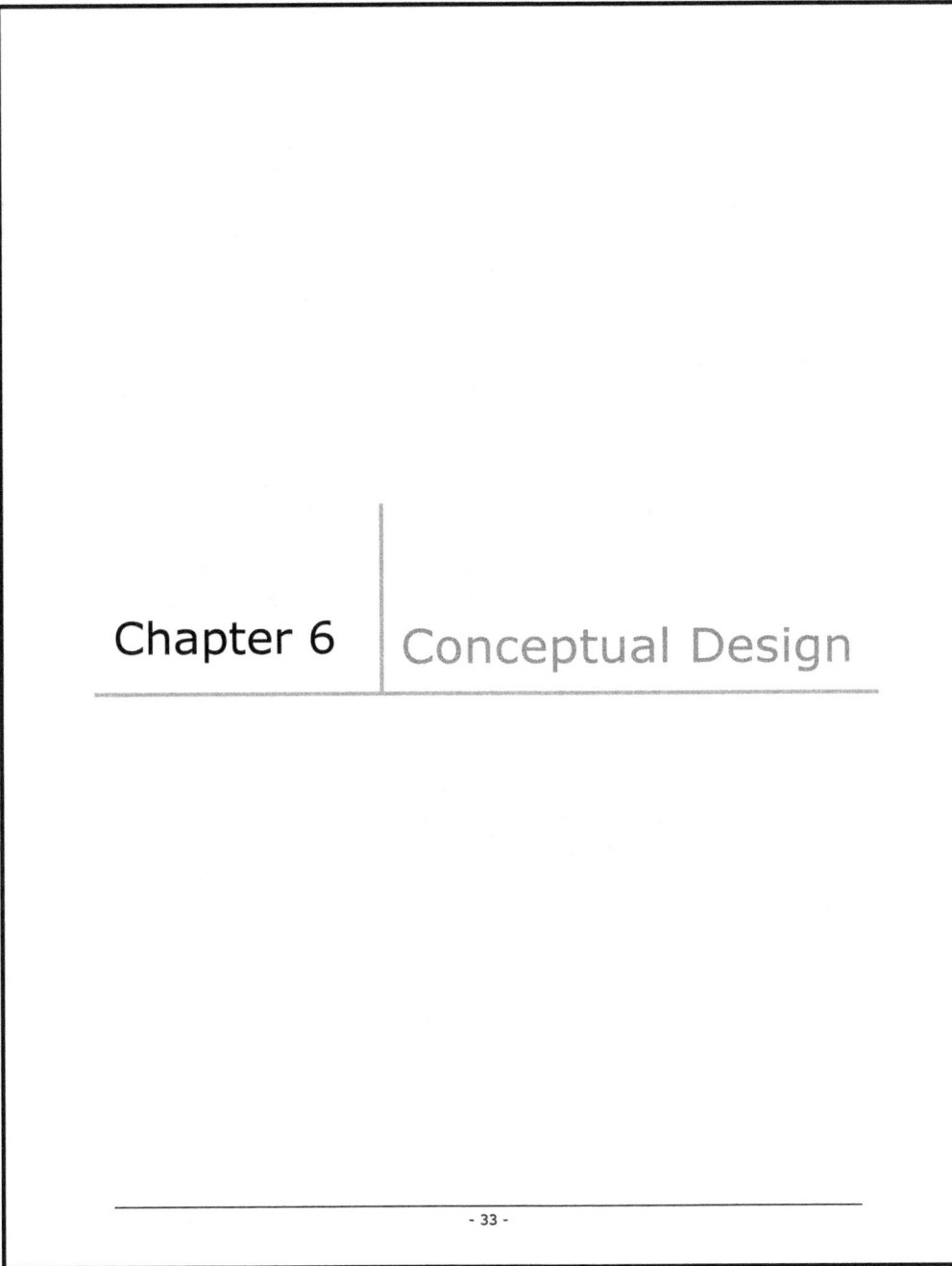

Chapter 6 Conceptual Design

6.0 Conceptual Design

Conceptual design phase of our aircraft design begins with the selection, comparison and evaluation of different reference airplane configurations for high altitude long endurance performance and the governing design parameters responsible for making aircraft to fulfill required performance and mission criteria. So, after analyzing different aircraft configurations many configurations were qualitatively eliminated based on feasibility considerations. The remaining configurations were analyzed based on figures of merit established after analyzing requirements. The mission was analyzed to establish the best configuration selection criteria initially. Furthermore, preliminary sizing, aerodynamic analysis and performance evaluation will validate the conceptual configuration selected initially.

6.1 Design Requirement Analysis

Design requirement analysis involves the detailed overview of design proposal and requirement for our particular aircraft. As high altitude and long endurance aircraft with solar powered mechanism have multiple applications so analyzing the history and already existing aircrafts and most importantly the available solar panels technology different figures of merit are evaluated for the selection of conceptual configuration in light of detailed design requirements.

Mission Requirements

The mission requirement involves the mission phases that the aircraft would be performing during its flight. A mission profile summarizes these phases. The mission profile of SAIST is shown below showing the major phases of its mission. The mission profile is simple having Cruise and loiter as the main operating segments although the mission involves specific sequence of takeoff and landing segments particularly for HALE UAV.

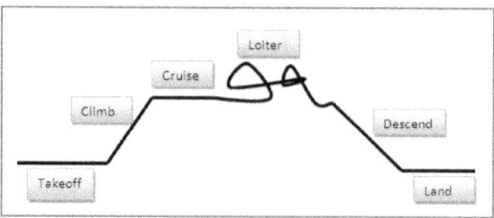

[Fig 6.1: Mission Profile SAIST]

Payload Requirements

Payload requirement majorly depends upon the applications of SAIST which are remote sensing, surveillance, local communication coverage and atmospheric study. So the payload may include the communication equipments, cameras, avionics and subsystem etc. After evaluating different solar powered aircrafts payload distribution and available application subsystems, the payload requirement of 100lbs was set initially for aircraft.

Propulsion System Requirements

Propulsion system requirements are very critical in design requirement. The important consideration in propulsion system requirement is the operation of HALE UAV and the mission profile in which most important design objectives are the range and endurance considerations. Thus continuous flight is possible using solar power system and an electrical propulsion system so that desired endurance and range is acquired during different applications. Final propulsion system has been selected after the completion of design process that evaluates the power requirements in the light of the design proposal or requirements. Optimization techniques will be applied in order to converge on the best propulsion system combination.

Solar Energy Requirements

Main power source for propulsion system is the solar energy thus how much solar energy is available for the propulsion system is also an important consideration in order to make continuous flight for required endurance. Solar panels efficiency is also a vital consideration in power system design. Power required for the propulsion system will evaluate the solar energy required from solar panels to the propulsion system.

Structural Requirements

The main objectives of SAIST suggests a light weight structure capable of producing high lift and low drag having strength and load handling capabilities enough to operate at high altitudes. Structural strength plays an important role in takeoff and landing sequences.

Altitude and Endurance Requirements

Altitude and endurance requirements are the basic and primarily objectives of SAIST. As per requirement an altitude of 80,000ft and an endurance of 12hrs have been proposed which will be attained during the design process and aircraft performance is made to converge at the best to attain the main design objective.

HALE observation Requirements

This requirement is basically from an application point of view and is mainly a function of altitude and formation of flying vehicle which will be accomplished in design as to cover the maximum area of observation and communication range as well.

Takeoff and Landing Requirements

Takeoff and landing is always crucial in unmanned remote controlled aircrafts. Operating speeds and stalling speed limits will be well defined for meeting the takeoff and landing requirements. Most importantly, proper landing gears or launching profile for takeoff in the absence of landing gears will be decided on the basis of design analysis. Remote handling of controls is also an important consideration for proper takeoff and landing.

General Aircraft Requirements

From requirement considerations and different configurations comparison it is concluded that SAIST HALE UAV must have the following features:

- High Aspect Ratio wing
- Flying wing configuration
- High endurance and range
- High altitude
- Solar powered
- Light weight structure
- Propeller driven
- Subsonic flight
- Stable flight

The requirements established after the analysis of HALE flying vehicle for our particular solar powered vehicle are:

- Aircraft should be capable of carrying payload of about 100lb.
- Aircraft should be capable to reconfigure the systems according to mission requirements.
- Ferry range must be 1500km = 810nm.
- Aircraft should have endurance of about 12hrs.
- All weather flight capability (day/night).
- Aircraft should be capable to takeoff and land with no specific systems. It may be short or on rough field or it may be a catapult/rocket launch, arresting wires with no specific recovery system.
- Aircraft should be capable of withstanding an appropriate structural loading during continuous flight.
- There should be a safety system to avoid or reduce aircraft failure.
- Aircraft should attain an altitude of 80,000ft.

6.2 Figures of Merit (FOM)

The figures of merit are established after analyzing design requirements in details. These FOMs will act as selection criteria for the best configuration as per design objectives.

These FOMs are as follows:

- Aerodynamic Efficiency
- Stability and Control
- Endurance
- Altitude
- Solar Energy
- Structural Integrity
- Weight
- Payload Handling
- Manufacturing
- Cost

Aerodynamic Efficiency

Aerodynamic efficiency is an important FOM, as the high drag and less aerodynamic configuration can severely affect the performance of the design.

Stability and Control

Stability and control in this HALE UAV is a very important figure of merit. Therefore, different configurations are compared to acquire longitudinal, lateral and directional stability and highest score will be given to that particular configuration which will be inherently stable. Adequate pitch, roll and yaw control without lift penalties is also considered in configuration comparisons. Furthermore, adequate lateral and directional control authority to allow the airplane to take off and land in a crosswind.

Endurance

It is an important FOM in determining the configuration of the aircraft that gives the highest endurance factor.

Altitude

High Altitude configuration is also one of the main FOM as this FOM is a function of aerodynamic efficiency, structural integrity and power source.

Solar Energy

The attainment of maximum solar energy is also a design consideration. Therefore, comparison is made to converge on the best configuration capable of absorbing solar energy through high efficiency solar panels.

Structural Integrity

During the manufacturing of a new design its structural integrity is very crucial. In order to attain the design requirement, a configuration selection is made which would be inherently feasible to structurally integrate without complexity.

Weight

Weight is also one of the important FOM in configuration selection as less weight results in low power requirements and also favors in optimum lifting characteristics.

Payload Handling

The payload carrying and storage capability plays a vital role in configuration selection as SAIST HALE UAV is a wide application oriented aircraft. Therefore, payload carrying and handling capability enhances its application.

Manufacturing

Manufacturing FOM is also one of the important considerations in configuration selection. As complex configuration will result in manufacturing problems, simple configuration will be selected to favor efficient manufacturing.

Cost

Limited cost is a constraint in configuration selection so this FOM is analyzed for different configuration and economically acceptable design will be selected.

6.3 Conceptual Configuration Alternatives

The first configuration is the ideal conventional type having a fuselage, tail and wing combination with some taper ratio. Being the most common configuration it gives an advantage of conventional trend of flight performance and optimum lift. But for the requirement of high altitude and long endurance flight this configuration is not favorable enough to acquire adequate desired performance and

[Fig 6.2: Concept 1]

propulsion system requirements also don't favor accordingly as per mission requirement.

[Fig 6.4: Concept 3]

Conventional configuration with V-tail having no elevators and polyhedral tapered wing configuration-tail offers low interference drag but it induces stability problems

[Fig 6.3: Concept 2]

through the adverse roll yaw-coupling of the rudder and along with manufacturing complications. Polyhedral wing enables inherent stability but it also involves manufacturing difficulties.

The next configuration is a sweep wing with winglets and a canard surface and pusher configuration. The main advantage of the canard configuration is that it is designed to stall before the wing to provide a safety mechanism. However, this leads to longer takeoff distances since the wing cannot reach its maximum lift coefficient. While it increases lift and reduces

[Fig 6.5: Concept 4]

induced drag, the interference from a canard causes the wing to be less efficient and increases the surface drag on the wing resulting in a neutral score in drag. The control surface on the canard is also required to be large since it has a shorter moment arm than an aft tail.

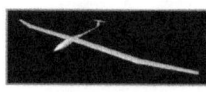

[Fig 6.7: Concept 6]

T-tail is heavier than a conventional tail due to the reinforcement necessary to support the horizontal stabilizer.

[Fig 6.6: Concept 5]

However, it allowed for a smaller horizontal stabilizer since it extended out of the wing wake and propeller wash. Moreover, a glider and seaplane type configuration enables a high glide ratio and aerodynamic efficiency due to smooth design but this configuration is not suitable due to drag penalty.

High aspect ratio wing enables good lift characteristics but twin rudder and blunt fuselage increases base and component drag although twin rudders are favorable in stability but coupling is also a problem in the particular configuration.

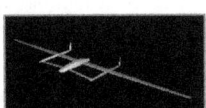

[Fig 6.8: Concept 7]

Two vertical stabilizers are mounted on either end of a horizontal tail gives vertical clearance benefits for comparable stability requirements. The horizontal surface is more efficiently loaded because the verticals surface act as winglets. But a fuselage twin boom gives drag penalty.

[Fig 6.9: Concept 8]

U-tail with swept wing at an incidence provides optimum performance but aerodynamically the configuration shown is not good enough to provide a high altitude and long endurance flight.

[Fig 6.10: Concept 9]

Joint tail, polyhedral wing and pusher configuration provides best performance at cruise conditions as best speed is attained with this configuration but drag values degrade the configuration for HALE performance. As low area coverage for solar panels and stability issues with the joint tail puts adverse effects on performance.

[Fig 6.11: Concept 10]

[Fig 6.12: Concept 11]

Delta wing configuration with the vertical stabilizers gives an outstanding performance but at high speed which is not the case for HALE solar aircrafts. So this configuration is completely ignored in the consideration of solar powered HALE vehicle.

Blended Wing Body

Blended Wing Body designates an alternative airframe design which incorporates design features from both a traditional tube and wing design into a hybrid flying wing configuration. The claimed advantages (as described below) of the BWB approach are efficient high-lift wings and a wide airfoil-shaped body. This enables the entire craft to contribute to lift generation with the result of potentially increased fuel economy.

[Fig 6.13: Concept 12] [Fig 6.14: Concept 13] [Fig 6.15: Concept 14]

In blended wing body configuration the control surfaces used are along with the wing as shown in the above figures dual vertical surface is placed close to tip or in the centre of the configuration and a combination of elevators and ailerons i.e. elevons are used on the wing for controlling of the aircraft. The aerodynamic characteristics and performance evaluation of blended wing are almost similar to flying wing configuration the only difference is in the drag values as presence of wing blended fuselage body although provides a space for payload placement but in addition to it a drag penalty also comes with it and interference drag will also be never zero.

Flying Wing

The continuous efforts for lower and lower drag led to the idea of the flying wing design. This offered the complete elimination of the parasite drag in addition to lighter weight and lower cost. But, at the same time, numerous problems of stability and control were to be overcome.

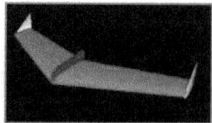

[Fig 6.16: Concept 15]

Flying wing is the most efficient aircraft configuration from the point of view of aerodynamics and structural weight. As absence of any aircraft components other than the wing should naturally provide many benefits. However, in practice an aircraft's wing must provide for flight stability and control; this imposes constraints on

[Fig 6.17: Concept 16]

the aircraft design problem. But gains in weight and drag reduction provides better efficiency and performance.

6.4 Detailed Comparison of Flying Wing Configurations

After analyzing different configurations and each and every possibility it is concluded that a flying wing configuration without twist is the most favorable configuration for our aircraft so different wing configurations in reference to "*NASA's research on HALE UAV analysis of alternatives*" are compared, keeping in mind the power collection and available surface area for solar panels final aircraft configuration according to our design requirement has been selected.

Figures of Merit established are the basis of comparison.

6.5 Initial Configuration Selection

6.5.1 Simple Rectangular Wing Configuration

The all-wing design has heritage in the family of solar-electric aircraft built by AeroVironment before and during NASA's ERAST program (Pathfinder, Pathfinder Plus, and Helios). All of these vehicles utilized distributed electric propulsion systems with numerous propellers driven by electric motors.

This configuration acts as a baseline configuration and the first most choice for solar powered HALE unmanned aerial vehicle.

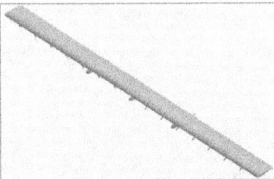

[Fig 6.18: Simple Rectangular Wing Configuration]

Some of the advantages and disadvantages of this configuration are as follows:

Advantages

- Flying wing without twist and sweep will result in inherent stability if reflex cambered airfoil is used, which is basically a flying wing airfoil.
- Tailless configuration and absence of fuselage will result a prominent decrease in the drag and it is one of the most important consideration and advantage of this configuration.
- All wing configurations with distributed propulsion system will result in steady and uniform flight.
- Solar panel efficiency is maximum and solar energy absorption is favorable due to exposed maximum surface area of the wing to the solar radiation.
- According to requirement a combination of sweep and twist in this configuration will result a particular stability.
- Due to symmetry of the configuration flying profile will be favorable in High altitude and long endurance flight.

Disadvantages

- Symmetry of the configuration will result in manufacturing complexes as a small fabrication error will result in stability problems during flight.
- In the absence of tail yawing moments that will effect flight.
- As this configuration is flying wing thus there is a low space for payload and subsystems placement within the wing.

6.5.2 Tapered Wing with dual surface Configuration

Tapered wing configuration with dual surface is also one of the considerations for HALE UAV. It's a flying wing configuration with distributed propulsion devices and bottom surface in order to increase the area for solar panels. Further, dihedral at tips favor in lateral stability.

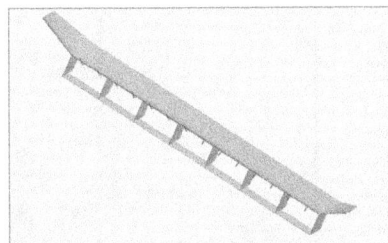

[Fig 6.19: Tapered wing with dual surface configuration]

Some of the advantages and disadvantages of this configuration are as follows:

Advantages

- This configuration has more surfaces available for solar panels and the vertical surface across wing and bottom surface provides more payload placement space.
- Dihedral at tips favor in lateral stability due to yaw motion.
- Distributed propulsion system will result in uniform and steady flight.

Disadvantages

- Due to more extra surfaces other than wing results in higher drag which is the most critical penalty for HALE UAV design motivation.
- Structural weight is greater which reduces useful payload weight according to our requirement.
- Due to higher drag more power is required so, more costly propulsion system is required for this configuration.
- Construction complexity is also a part of this configuration.

6.5.3 Joint Wing Configuration with Sweep

Joint wing configuration is famous for its high speed flight. The configuration shown is one o the unconventional configurations but considered for HALE UAVs alternative as well due to some advantages regarding solar energy benefits.

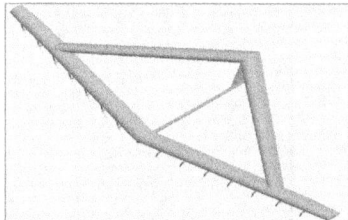

[Fig 6.20: Joint wing Configuration]

Some of the advantages and disadvantages of this configuration are as follows:

Advantages

- More surface area for solar panels so more power attainment for propulsion system.
- Rudder control surface favors in aircraft control.
- Swept wing favors stability and flight speed.
- Distributed propulsion system will result in uniform and steady flight profile.

Disadvantages

- Due to joint surfaces there exist interference drags that will result in increment in overall drag.
- More Power requirement due to higher drag.
- Low aerodynamic efficiency

6.5.4 Straight Wing Configuration with extra Surfaces

This configuration is particularly designed in accordance with solar panels efficiency and solar powered propulsion system. Movable surfaces along the wing with the solar panels will result in more solar energy absorption by orienting them along the maximum solar radiation during flight.

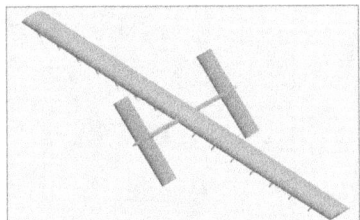

[Fig 6.21: Straight wing configuration with movable surfaces]

Some of the advantages and disadvantages of this configuration are as follows;

Advantages

- Solar panels become effective by using movable surfaces with solar panels mounted on it.
- The configuration contains solar panels fixed on wing as well as movable solar panels.
- Distributed propulsion system will result in steady and uniform flight profile.

Disadvantages

- Lower aerodynamic efficiency and there exist turbulent flow at the leading edge of the configuration in the presence of movable surface.
- Interference drag also reduces performance.
- Specific control mechanism will be required for movable solar panels in order to align their orientation according to solar radiation.

6.5.5 Figures of Merit Comparison

Summary of FOMs comparison for above four configurations is as below:

FOMs	Weights	Configurations			
Aerodynamic Efficiency	5	4	3	2	2
Stability and Control	4	2	3	3	2
Endurance	5	4	2	2	3
Altitude	5	4	2	2	3
Solar Energy	3	2	3	3	3
Structural Integrity	3	3	2	1	3
Weight	3	3	2	1	2
Payload Handling	2	1	2	1	1
Manufacturing	3	3	2	1	1
Cost	2	2	1	1	1
Total	35	28	22	17	21

[Table 6.1: Configuration Comparison]

Finally, figures of merit comparison shows the maximum points for simple rectangular wing configuration.

So, initially simple rectangular wing with polyhedral and distributed propulsion system has been selected.

6.6 Final Initial Conceptual Configuration Selected

So, after detailed analysis of mission requirement and desired performance and also the case study of NASA's Helios configuration finally the configuration selected is the rectangular wing with no sweep and no twist which inherently favors in stability and moreover, due to absence of any additional component besides wing, the drag values are sufficiently reduced that enables lower power consumption and lower thrust required values. As solar powered propulsion system includes the placement of solar panels and flying wing will provide all the available area and favorable solar radiation absorption through solar panels as maximum area is exposed to the sun.

[Fig 6.22: Artistic sketch- Initial conceptual configuration selected]

6.7 Drag Contribution Comparison

The comparison of configuration with the conventional one enables us to better evaluate the flying wing shown above. The distribution of drag which is the most important factor for better performance of solar powered HALE flying vehicle is as;

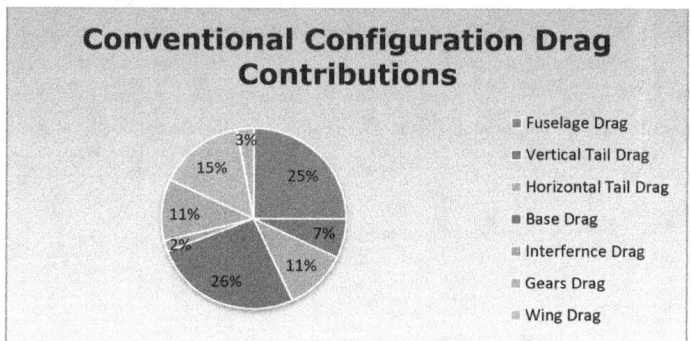

[Fig 6.23: Conventional Configuration Drag Distribution]

Reduced drag contribution for flying wing is shown below. The absence of fuselage, tail and gears enable reduction in the form and skin friction drags associated with those components and the only and main contribution in drag is that of flying wing.

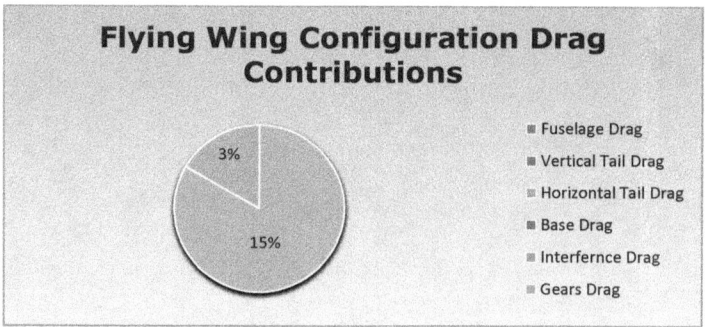

[Fig 6.24: Flying Wing Configuration drags Contribution]

It seems that drag reduces to about **82 %** from conventional configuration.

Chapter 7

HALE Earth Observation

7.0 HALE Earth Observation Formation

As the Aircraft has to operate at high altitudes, the considerations of earth observation are also considered accordingly. From geometry considerations it is known that the maximum earth observation range is a function of height above sea level and the radius of earth. Moreover, earth is not purely a sphere, being flattened at the Polar Regions relative to the equator. The exact earth radius is 6356.9km at the poles and at the equator the radius is 6378.4km.For analyzing a mean value of 6378.1km is considered.

When we are flying, if we look down below the horizon the observation range becomes shorter and the view is clearer. This improved vision is partly due to reduced atmospheric contamination and ground clutter that adversely affect longer viewing distances. Simple trigonometry can be used to determine the observation range at various altitudes and for different downward-viewing angles (slant angles).

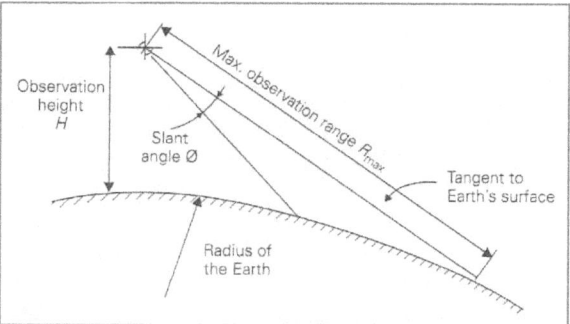

[Fig 7.1: Observation Geometry]

At an altitude of 20 km (65000ft), the horizon (zero slant angles) is about 500km (270nm) away. Looking down 5 degrees, the distance to the ground reduces to about 200km (108nm).

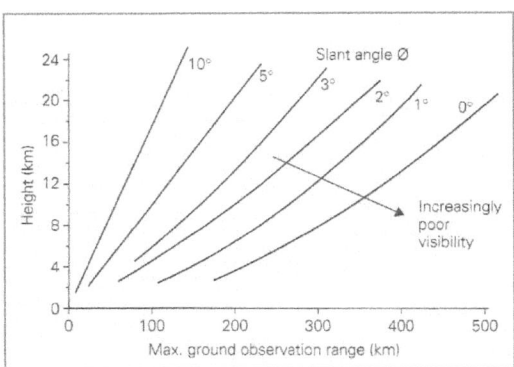

[Fig 7.2: Altitude versus observation range at different slant angles]

The other factor in observation range is the reconnaissance equipment of the aircraft. An efficient system gives an advantage to operate at high altitude to observe a greater surface area. In order to cover a wide range of area, the airplane has to fly at a high altitude to cover the required surveillance area.

For an aircraft capable of continuous flight, all weather operation is another very important consideration. For this purpose, wind speed at operating altitude is studied. If wind speed is constant then the aerodynamic effects are less significant on the aircraft. Where, in the case when wind speed is not constant the overall performance and aerodynamic efficiency varies relative to wind speed.

7.1 Wind Speed Variation with Altitude

The variation of the wind speed with altitude leads to the generation of vertical wing loading on the aircraft (gust effects).Although gusts are less pronounced at higher altitudes they must still be considered. Such gust activity leads to dynamic changes in aircraft lift. Aircraft with low wing loading or high lift curve slope are more susceptible to gust disturbance. Aircraft operating at high altitudes are designed as to count for all aerodynamic effects.

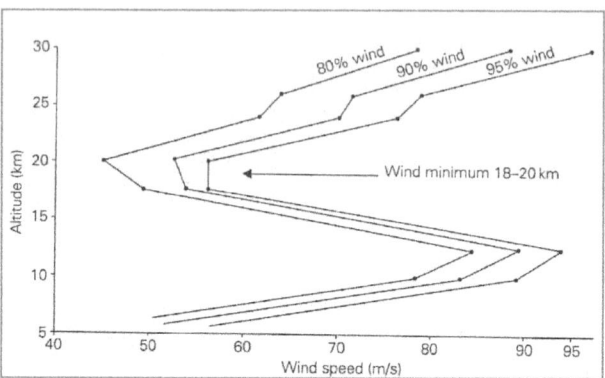

[Fig 7.3: Wind speed variation versus altitude]

7.2 Temperature variation with altitude

The other influencing factor is the temperature variation along the changing altitude but initially it is assumed that the temperature has no effects on the material and aerodynamics of the aircraft.

HALE aircrafts flying zone lies in stratosphere and particularly for SAIST, according to the requirement, has flying altitude of about 24km and the temperature variation along that region is shown in the figure below:

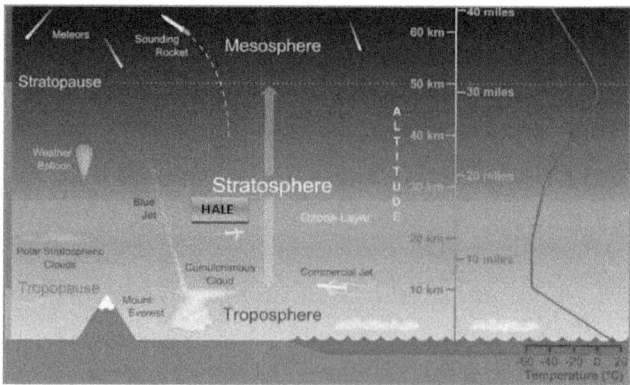

[Fig 7.4: Altitude versus temperature variation]

7.3 HALE Observation Formation

So after analyzing high altitude characteristics and our aircraft design requirements initially SAIST flying orientation and zone is defined as:

- Slant Angle: 1deg
- Altitude=80,000ft/24km
- Max Ground Observation Range: 400km(Ground curvature)

Chapter 8 | Preliminary Design

8.0 Preliminary Sizing

Preliminary design phase started after the finalization of aircraft configuration. In this section sizing, airfoil selection, aerodynamics, performance, stability and propulsion estimation has been made. After the finalization of geometry a 3D CAD model has been developed. Preliminary design further leads to analysis phase in which iteration of design parameter results in convergence of optimized design.

8.1 Initial Weight Estimation

Weight estimation is the most important and critical part of the preliminary design. In this case of solar powered HALE UAV, weight has been estimated using statistical data and trends from historical solar powered aircrafts developed earlier. Initial weight estimation is then verified and iterated to converge on the best configuration geometry fulfilling design requirement.

Weight estimation influence overall design of an aircraft so, initial estimate of SAIST is made from reference weights distribution.

We are designing a solar powered propeller driven electric powered flying vehicle so almost all weights are fixed as there is no payload drop or fuel consumption.

Total Weight: 300 lbs (Initial Estimate in reference to requirement)

Weight Division

After case studies and specification analysis weight distribution is as follows:

- Airframe structure: **18%**
- Propulsion system: **12%**
- Solar powered mechanism equipment: **23%**
- Payload: **35%**
- Control Equipment: **5%**
- Avionics system: **3%**
- Miscellaneous (Landing gears, nuts, epoxies, mounts etc): **4%**

Initial Wing Loading Selection

In order to fly at high altitudes the configuration selected in conceptual phase must have more wing area than weight so from reference data and case studies the wing loading selected in this case is **0.568 lb/ft²**.

8.2 Initial Geometry Sizing

Wing loading has been estimated so geometry parameters are calculated as:

$$W/_S = 0.568 \ lb/ft^2$$

$$W=300lbs$$

As according to requirement our design must have high aspect ratio in order to have better lift so aspect ratio of 14 has been selected.

$$AR=14$$

$$AR = \frac{b^2}{S}$$

Wing Area comes out to be, S=528ft²

Wing span, b=85.97ft=86ft

As in our case it's a rectangular wing having taper ratio of 1 so root and tip chord comes out to be,

$$C_{root}=C_{tip}=2S/b \ (1+\lambda) =6.139ft$$

Aerodynamic center=1.534 ft (0.25C)

Reynolds Number Initial Calculation

Reference Reynolds number at cruise conditions is calculated as;

Initial Design altitude=40,000ft

Initial Design Velocity=90ft/sec

$$Re = \frac{\rho V c}{\mu}$$

Where,

$$\rho = \text{Density at 40,000ft} = 0.00058727 \text{ slug/ft}^3$$

$$V = \text{Cruise velocity} = 90 \text{ ft/s}$$

$$c = \text{Chord length} = 6.139 \text{ ft}$$

$$\mu = \text{Coefficient of viscosity of air} = 1.7894 \times 10^{-5} \text{ kg/ms}$$

$$Re = 8.5 \times 10^5 \text{ approx. (this result will be used for airfoil selection.)}$$

Initial Design Lift Coefficient Calculation

In order to calculate deign lift coefficient we first calculate dynamic pressure as:

$$q = \frac{1}{2}\rho V^2 = 2.378 \text{lb/ft}^2$$

$$C_L = \frac{W/_S}{q}$$

$$C_L = 0.238 \ (\textit{Too low for better performance})$$

For Pathfinder (Reference values)

The NASA's pathfinder HALE aircraft acts as a reference configuration which has the following characteristics;

- $W = 560\text{lb}$
- $S = 792\text{ft}^2$
- $C_{L \ max} = 1.3$
- $V_{stall)\ Pathfinder} = 69.7\text{ft/sec}$

Power to weight ratio (Hp/lb) Estimation

Power to weight has been estimated from empirical formula suggested by Raymer as:

$$P/W_0 = a V_{max}{}^c$$

For Home built Composites a=0.004, c=0.57

$$V_{max} = 95 \text{ft/sec}$$

$$P/W_o = 0.0536 \text{hp/lb}$$

Power Loading

Power loading comes out to be:

$$W/P = 18.648 lb/hp$$

8.2.1 Wing Loading Calculations

Wing loading for different segments of the flight profile has been estimated keeping in mind the initial design requirements of altitude of 40,000ft.

Stall

Wing loading during stall is estimated by relation suggested by Raymer as:

$$W/S \leq \frac{1}{2}\rho V^2 C_{Lmax}$$

$$W/S \leq \frac{1}{2}(0.5861 \times 10^{-3})(69.7)^2(1.3)$$

$$\mathbf{W/S \leq 1.84}$$

Take-off

Wing loading during takeoff comes out to be:

$$TOP = 70$$

$$CL_{T.O} = CL_{max} (V_{stall}/V_{Take\ off})^2$$

$$CL_{T.O} = 1.3 (1/1.1)^2$$

$$CL_{T.O} = 1.07$$

$$\frac{W}{S} = (TOP)\sigma CL_{T.O}\left(\frac{P}{W_o}\right)$$

$$\sigma = 1$$

$$(P/W_o)_{max} = 0.0804 \text{ (At take off)}$$

$$W/S = 70 \times 1 \times 1.07 \times 0.08045$$

$$\frac{W}{S} = 6.02196 \text{lb/ft}^2$$

Thrust to weight Ratio during Climb

Thrust to weight ratio calculation during climb is calculated as:

$$\frac{P}{W} = \frac{0.043\,hp}{lb} \text{(From table 5.4)}$$

$$\frac{T}{W} = \left(\frac{550\eta_p}{V}\right)\frac{P}{W}$$

$$\frac{T}{W} = 0.184$$

$$q = \frac{1}{2}\rho V^2 = 0.906 lb/ft^2$$

Wing Loading at Cruise

Wing loading for cruise segment of mission profile is:

$$\frac{W}{S} = q\sqrt{Ae\pi C_{DO}}$$

Initially estimated $C_{DO} = 0.013$

$$e = 0.8$$

$$\frac{W}{S} = 0.6127$$

Wing Loading for Loiter Endurance

Wing loading during loiter segment is calculated as:

$$\frac{W}{S} = q\sqrt{3Ae\pi C_{DO}}$$

$$\frac{W}{S} = 1.06$$

Wing Loading for Landing

Wing loading estimation during landing is as:

$$(S_{\text{landing}})_{\text{Pathfinder}} = 453.15\text{ft(Using reference value initially)}$$

$$\sigma = 0.0002238/0.002377 = 0.0941$$

$$S_{\text{landing}} \leq 500\text{ft (including Sa)}$$

$$S_{\text{landing}} = 80 \left(\frac{W}{S}\right)(1/C_{\text{Lmax}})$$

$$\frac{W}{S} = 0.76$$

Wing Loading during sustained Turn

Wing loading during sustained turn is calculated as:

$$W/S = 0.078$$

Wing Loading for maximum ceiling

Wing loading for maximum ceiling is calculated as:

$$\frac{W}{S} = q\sqrt{\frac{Ae\pi C_{DO}}{3}}$$

$$\frac{W}{S} = 0.30653$$

8.2.2 Summary of Wing Loadings

Segment	Wing Loading
Take-off	6.021
Cruise	0.612(Selected)
Loiter	1.06
Stall	1.84
Sustained Turn	0.078
Maximum ceiling	0.306
Landing	0.76

[Table 8.1: Wing Loading For different Segments]

Selected Wing Loading

After analyzing all segments particular aircraft requirements have been finally selected the following wing loading as a design point.

$$\frac{W}{S} = 0.612$$

8.2.3 Thrust to weight ratio and power loading

So, after the finalization of wing loading modified thrust to weight ratio and power loading which are the function of wing loading are modified as:

$$T/W = 0.242 \ (From \ Equation \ 5.24)$$

$$n = 3.2 (Initial \ value)$$

At this value of thrust loading the power loading comes out to be;

$$P/W = 0.0561$$

Initial Empty Weight Fraction

Initial empty weight fraction for SAIST comes out to be;

$$We/Wo = 0.6466$$

8.3 Modified Sizing and Design Considerations

After selecting a modified wing loading almost all design parameters are changed as;

$$W/S = 0.612$$

$$AR = 14$$

$$W = 323.136lb$$

Modified parameters are thus

$$S = 528ft^2$$

$$AR = b^2/S$$

$$b = 85.97ft = 86\ ft$$

As the wing is straight,

$$\text{Taper ratio} = \lambda = 1$$

$$C_{root} = \frac{2S}{b(1+\lambda)} = 6.139ft$$

$$C_{tip} = \lambda\, Croot = 6.139ft$$

Modified Design Altitude=60,000ft (As conditions are more favorable at this altitude)

8.3.1 Modified Design Lift Coefficient

So, after refining sizing and modified design altitude coefficient of lift is calculated as:

$$C_L = \frac{W/_S}{q}$$

$$q = \frac{1}{2}\rho V^2 = 0.891 \text{lb/ft}^2$$

$$V = 74 \text{ft/sec}$$

$$W/S = 0.612$$

$$\mathbf{C_L = 0.68}$$

The lift coefficient value may be modified during aerodynamic analysis as for best suited design velocity with required altitude.

This value of the design lift coefficient will be modified after the optimization of design speed.

Validation of Weight & Wing Loading through Trend

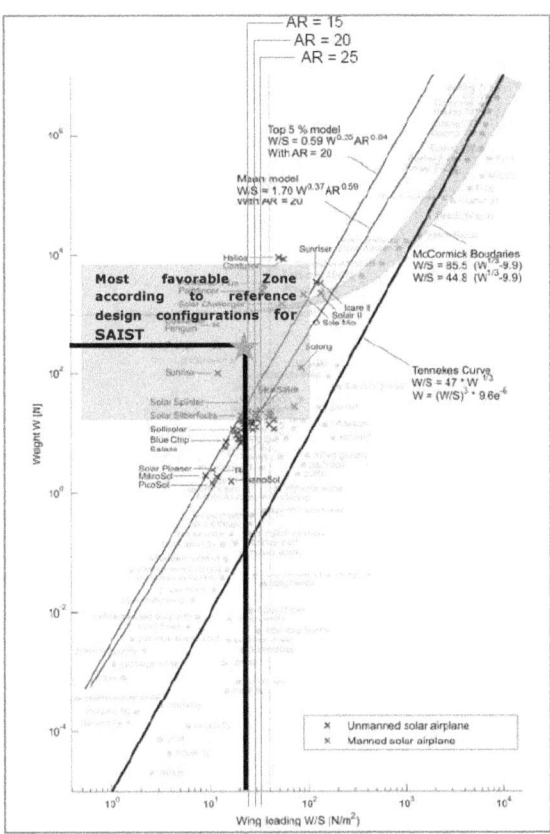

[Fig 8.1: Great Flight Diagram with 92 solar airplanes]

Thus trend verifies the close values with the reference aircraft configurations specifically Helios, Centurion, Pathfinder and Zephyr.

Thus,

$$W = 1436.13N = 323.13lbs, AR = 14$$

$$\frac{W}{S} = 29.77 \left(\frac{N}{m^2} \right) = 0.612 \text{lb/ft}^2$$

In case of sailplanes considerations the aircraft design value comes out to be as shown on great flight diagram completed with 415 sailplanes by Tennekes H (1996).

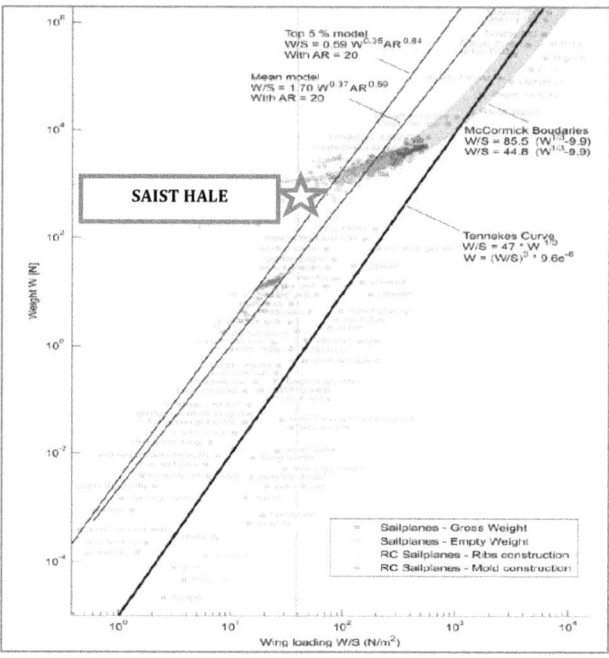

[Fig 8.2: Great Flight Diagram with 415 Sailplanes]

From the analysis of both great flight diagrams it is verified that SAIST also lies within solar airplanes and sailplanes trend thus geometry calculations along with weight and wing loading estimation is verified.

8.4 Airfoil Selection

The selection of airfoil for a fighter aircraft is very critical and before analyzing different airfoils of different reference aircraft, it's very important to analyze the estimated performance of the aircraft and for the selection of airfoil its more convenient to define the proper figures of merit that must be essentially fulfilled for our aircraft that has to perform a prescribed mission.

8.4.1 Airfoils Comparison

Airfoils	Shape	Cl_{alpha}	Cl_{max}	t/c	Stall AOA
MH 49		0.074	1.0303	0.01628	11.5
mhmi3		0.082	1.240	0.01562	11.5
MH 61		0.070	1.0070	0.01671	10
MH 46		0.081	1.173	0.01840	12.5
mhmi2		0.107	1.263	0.01510	11.5
MH 64		0.500	1.052	0.01400	10
MH-78		0.070	1.346	0.0235	13
MH-45		0.070	1.200	0.01602	11.5

[Table 8.2 Airfoils Comparison]

So the figures of merit in this regard are:

- Long Endurance
- High Altitude
- Structural Weight
- Lowest Drag
- Stable Flight
- Aerodynamic Efficiency
- Handling Qualities

- Take off & Landing Distance
- Maximum Range

Design Lift Coefficient

Initially design lift coefficient is calculated as

$$C_L = 0.68$$

8.4.2 Airfoil Comparison Plots

Airfoil coefficient of lift versus drag coefficient Plot:

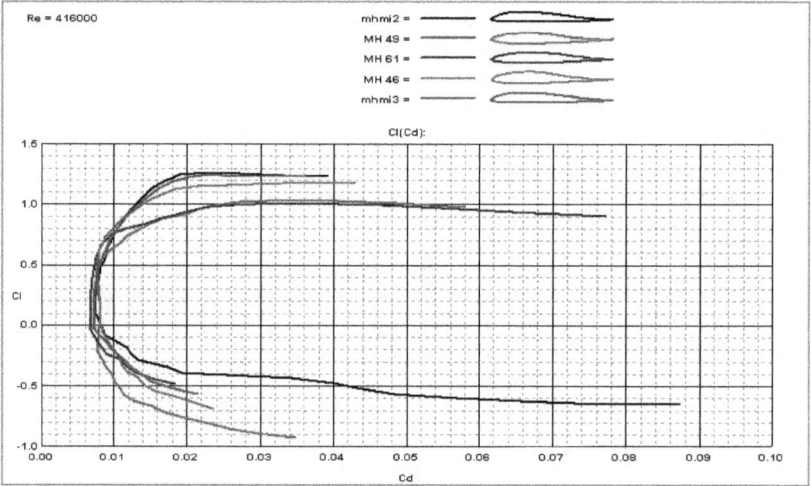

[Fig 8.3: Airfoils Lift Coefficient versus Drag Coefficient Plot]

The above plot shows lift coefficient variation with drag coefficient. The drag polar curves above for different airfoils that has matched the airplane requirements after studying reference and suggested airfoils in consideration of lift, endurance and stability concerns.

Airfoil Coefficient of Lift and drag coefficient Versus Angle of Attack

The lift coefficient variation with respect to angle of attack curves for different airfoils shows the maximum lift coefficient which also responds to stall angle. Moreover, the drag coefficient variation with the change in angle of attack provides us optimum values of drag coefficient at different conditions.

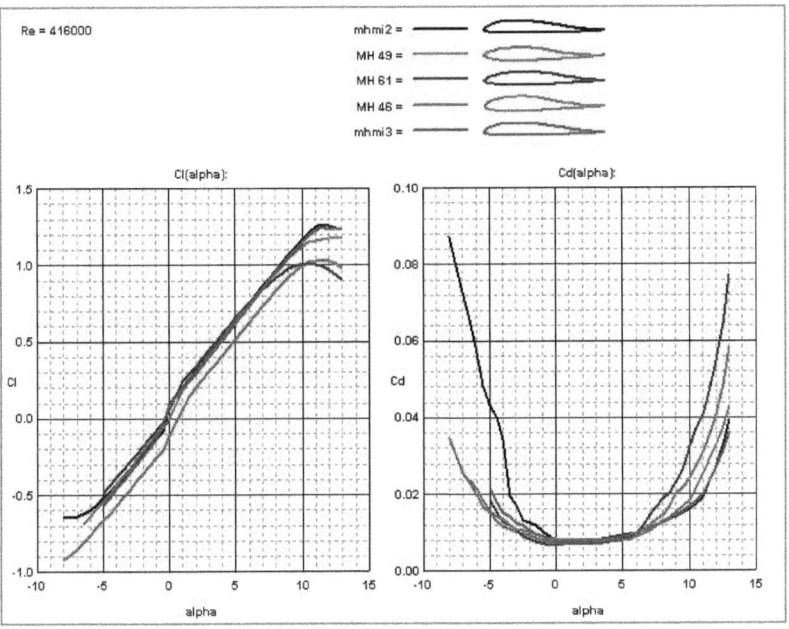

[Fig 8.4: Different airfoils Cl-alpha & Cd-alpha plots]

Airfoil Cl/Cd versus alpha and moment coefficient plots

Aerodynamic efficiency which is Cl/Cd ratio with respect to angle of attack defines the performance of the aircraft. Most importantly, moment coefficients at various angles of attacks for different airfoils are an important consideration for stability issues.

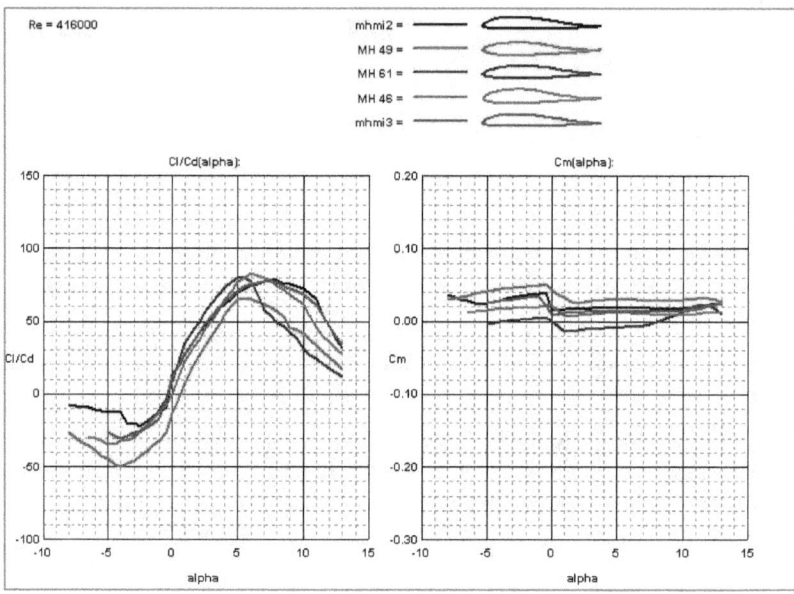

[Fig 8.5: Different Airfoils Cl/Cd & Cm versus alpha Plot]

Further Airfoils Comparison Plots:

Lift coefficient vs. drag coefficient plots

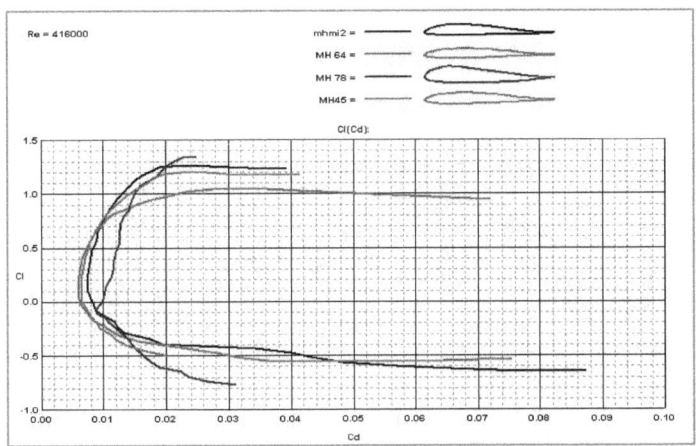

[Fig 8.6: Airfoils Cl-Cd Plots]

Lift Coefficient and Drag Coefficient vs. Alpha plot

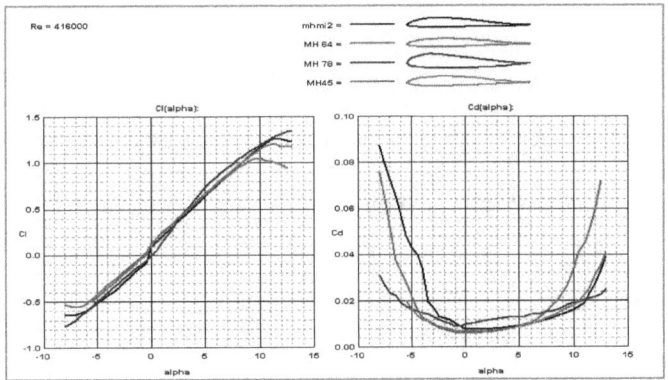

[Fig 8.7: Airfoil Comparison Plots]

Airfoil Cl/Cd versus Alpha and Moment Coefficient Plots

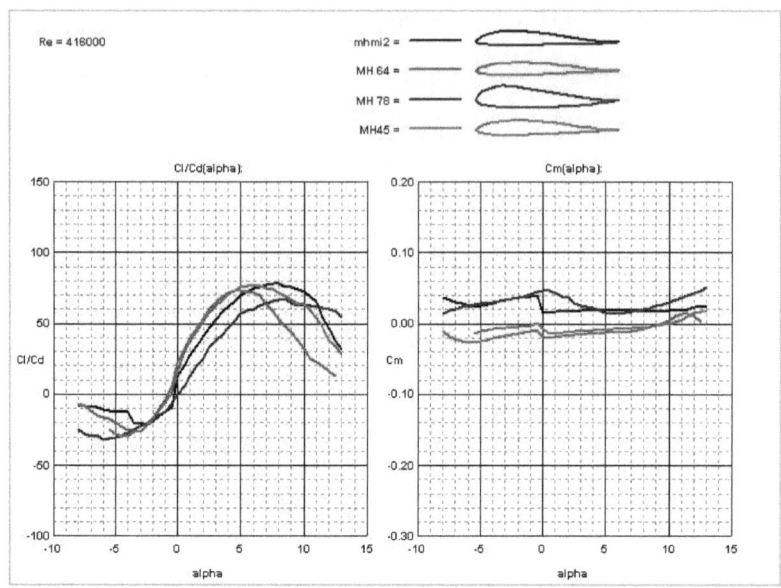

[Fig 8.8: Airfoil Comparison Plot Cl/Cd & Cm versus alpha plot]

8.4.3 Conclusion

After analyzing different airfoils and their plots, the best suited airfoil comes out to be **mhmi2** as it has the most favorable characteristics according to the requirements of the solar powered HALE airplane. In conceptual base configuration selection it has been justified to make an airplane with flying wing configuration as due to favor in drag reduction, high performance and best suited stability.

The reflexed cambered airfoil has been chosen after analyzing different airfoils.

Tailless planes and flying wings can be equipped with almost any airfoil, if sweep and twist distribution are chosen accordingly. However, if we want to design a tailless plane with a wide operating range, the wing should have a small amount of twist only, or none at all, to keep the induced drag at reasonable levels throughout the whole flight envelope. Under these conditions, the wing must not create a large variation in moment coefficient, when the angle of attack is varied. This makes it necessary, to use airfoils with a low moment coefficient. In the case of an unswept wing as initially in our design ("plank"), even an airfoil with a positive moment coefficient is necessary, to avoid upward deflected flaps under trimmed flight conditions. Such airfoils usually have a reflex camber line.

The horizontal tail plane provides the necessary amount of longitudinal stability on a conventional plane; it is the wing, which stabilizes an unswept wing. In most cases, airfoils with *reflex* (s-shaped) mean lines are used on flying wing models to achieve a longitudinally stable model.

In order to provide longitudinal stability to a wing, two things are important:

Total Force and Moment, c/4 point:

The pressure forces, which act on the surface of each wing section, can be replaced by a single total force and a single total moment. Both act at the *quarter-chord point* of the airfoil. When the angle of attack changes (e.g. due to a gust), the moment stays nearly constant, but the total force changes. Increasing the angle of attack increases the force.

Center of Gravity:

Translations and rotations of "free floating" bodies are performed relative to their center of gravity. When the angle of attack of a plane changes, the plane rotates (pitches) around its center of gravity (*c.g.*).

So a reflexed cambered airfoil with unswept and untwisted wing will provide a most favorable combination for optimum performance in our design of solar powered high altitude long endurance airplane.

8.4.4 Figures of Merit Comparison

The comparison of different airfoils is shown as below

Figures of Merit	Weight	MH-49	MH-61	MH-46	mhmi3	MH-64	MH-78	MH-45	mhmi2
Drag	4	3	2	2	2	2	3	2	3
Flight Performance	4	2	1	2	2	2	3	2	3
Stability and Control	4	1	2	2	4	1	1	1	4
Handling Qualities	3	1	2	2	2	1	2	2	2
Structural Weight	3	2	2	1	2	2	2	1	2
Aerodynamic Efficiency	3	3	1	3	2	2	2	1	2
Total	21	42	35	42	50	35	46	32	58

[Table 8.3: Airfoil Selection from FOMs Comparison]

So, finally FOM comparison chart for considered airfoils for SAIST is

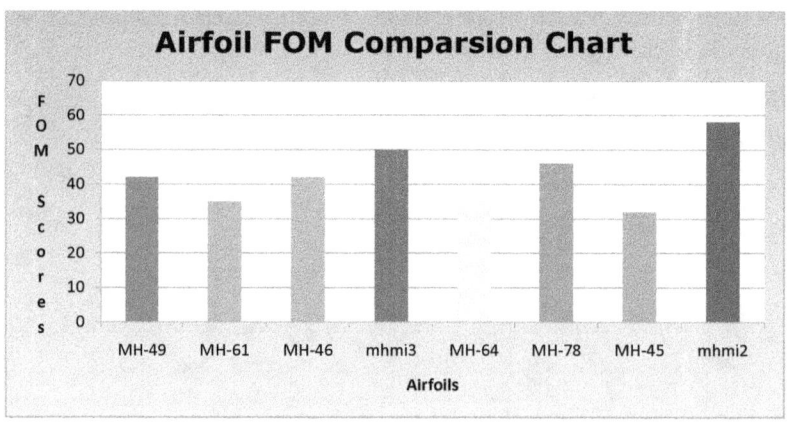

[Fig 8.4: Airfoils FOM Comparison Chart]

Best suited airfoil as already discussed above is **mhmi2**.

8.4.5 Airfoil Selection

[Fig 8.5: mhmi2 Airfoil]

This airfoil is taken from Profili software and is designed by an Italian engineer Matteo Gallizia

- Airfoil maximum thickness: 9.31% at 23.7% of the chord.
- Maximum camber: 2.4% at 27.2 %of chord
- Leading Edge Radius: 0.8483%
- Trailing edge thickness:0.0001%

8.4.5.1 Airfoil 'mhmi2' Plots Using Design Foil

The plot below shows the variation of lift coefficient with drag coefficient and also lift coefficient and moment coefficient variation along with angle of attack.

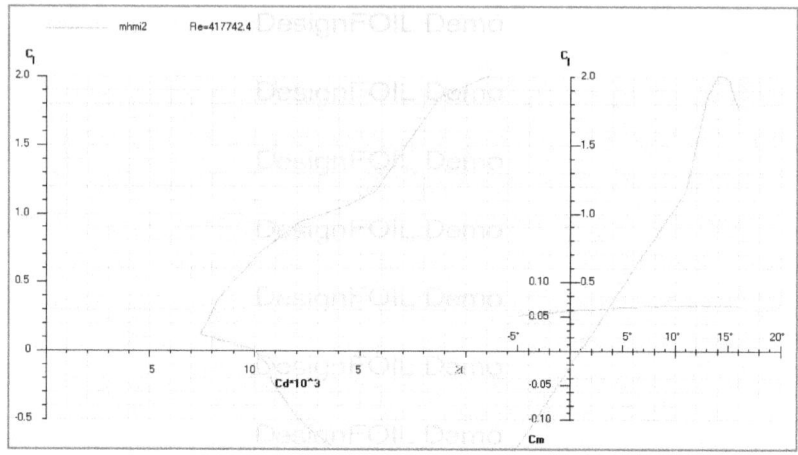

[Fig 8.6:mhmi2 Airfoil Cl-Cd & Cm, Cl versus alpha Plot]

Drag Coefficient versus Angle of attack Plot:

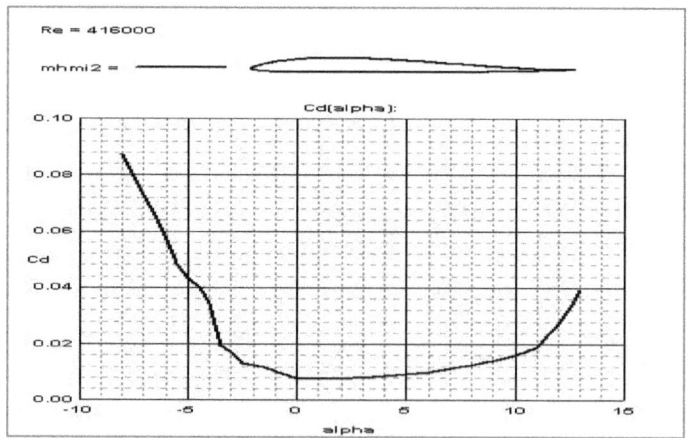

[Fig 8.7: mhmi2 Cd versus alpha plot]

Cl/Cd vs. Alpha Plot

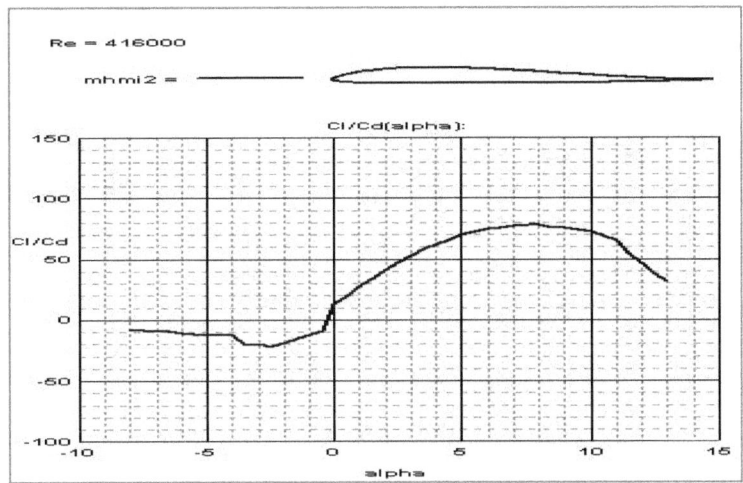

[Fig 8.8: Cl/Cd versus alpha plot for mhmi2 airfoil]

8.4.5.2 Pressure Distribution Curves

Airfoil pressure distribution along the airfoil at zero angle of attack is shown.

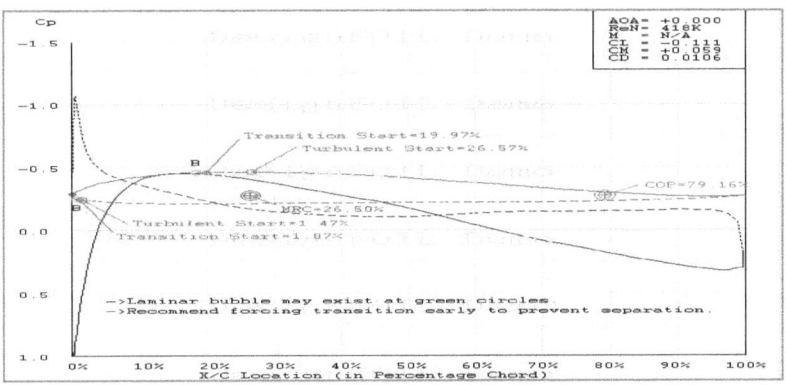

[Fig 8.9: Airfoil mhmi2 Cp Distribution at zero angle of attack]

The plot shows the transition start and turbulent flow originating as we move along the airfoil at zero angle of attack and it is showing the laminar bubble indication that results in increment in drag during steady flight at zero angle of attack.

Airfoil Flow speed variation at zero angle of attack

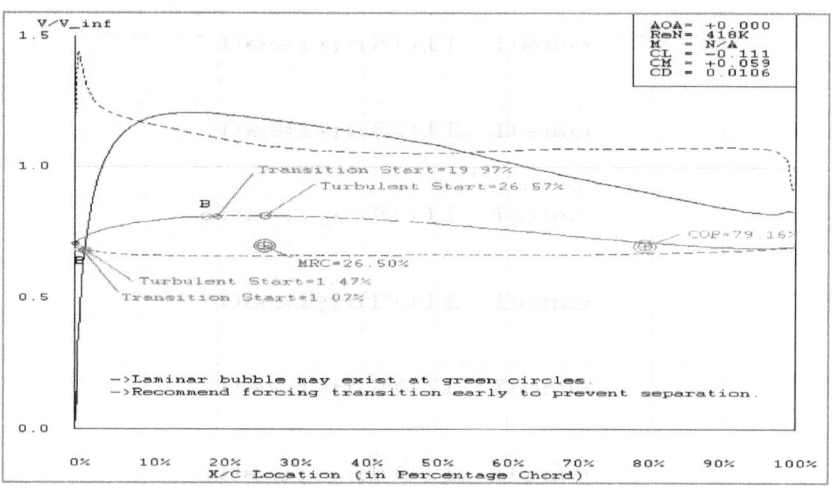

[Fig 8.10: Speed Variation along Airfoil mhmi2]

Boundary Layer Thickness Variation

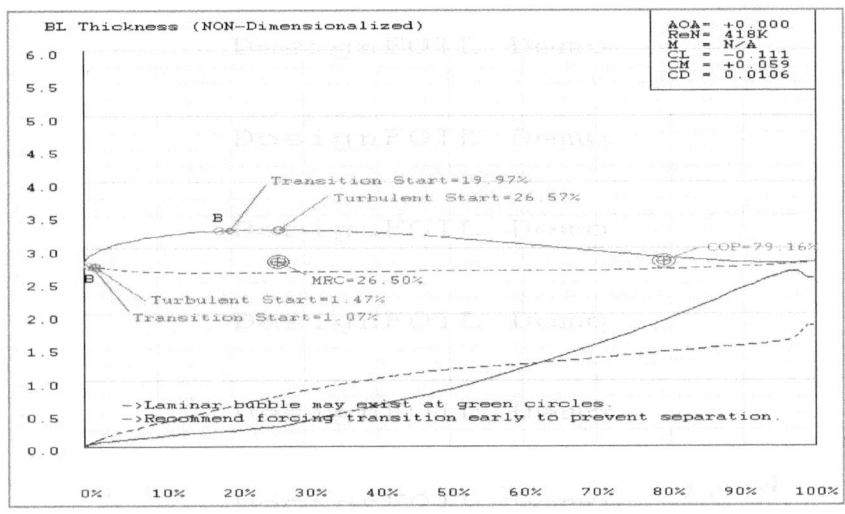

[Fig 8.11: Airfoil mhmi2 Boundary Layer thickness variation]

Chapter 9 | Aerodynamics

9.0 Aerodynamics

Aerodynamics involves the prediction of characteristics and behavior of flight. Specifically, in HALE UAVs design aerodynamics plays a major role in optimized design. Further, aircraft behavior at different altitude, its flight profile, drag variations and thrust requirements are the key design estimations.

Modified Reynolds Number

Reynolds number is calculated at 60,000ft as

$$Re = 4.16 \times 10^5$$

9.1 Lift Curve Slope

Airfoil Lift Curve Slope

Airfoil lift curve slope from airfoil analyzed data is as

$$C_{l,\alpha} = (0.2591 - 0.2054)/(1.5 - 1.0) = 0.1075/degrees$$

$$C_{l,\alpha} = 6.20275/radians$$

Aircraft Lift Curve Slope

Using the relation from reference for solar powered vehicles

$$C_{L,\alpha} = f(\frac{C_{l,\alpha}}{1 + \frac{C_{l,\alpha}}{\pi AR}}) = 5.32695/radians$$

$$f = 0.98$$

9.2 Drag Calculations

Drag is the most important parameter for SAIST as most of the performance and propulsion system requirements depend upon it. So, drag calculations have been made precisely and in detail considering origin of any possible drag in SAIST configuration.

Aircraft has divided zones as shown below

| Tip Panel | Side Panel | Mid Panel | Side Panel | Tip Panel |

[Fig 9.1: SAIST panels division]

Where dimensions of the zones are as follows:

- Mid Plane=19.48 ft
- Side Plane=16.65 ft
- Tip Plane=16.65 ft

9.2.1 Types of Drag

For the calculation of drag, the following are the types of drag encountered in the aircraft

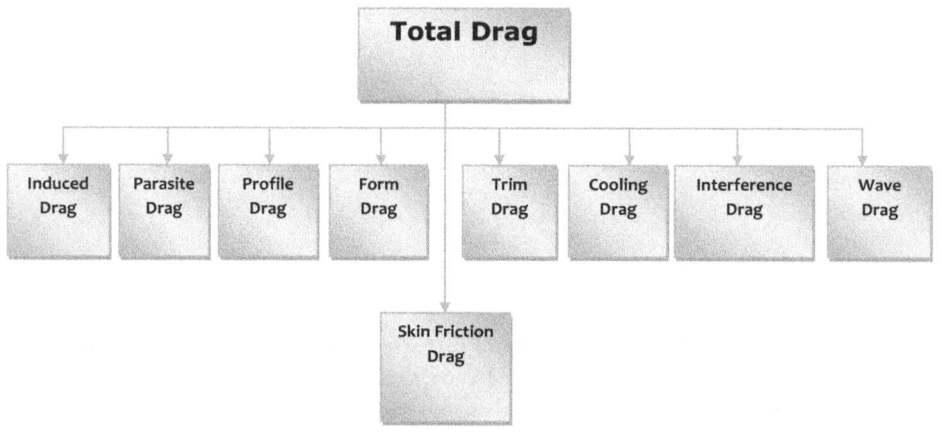

[Fig 9.2: Types of Drag]

9.2.1.1 Induced drag

The drag results from the generation of a trailing vortex system downstream of a lifting surface of finite aspect ratio.

$$CDi = K\,CL^2$$

$$K = 1/\pi eAR$$

$$e=0.65,\ AR=14$$

$$K=0.03499$$

$$CDi = 0.03499\,CL^2$$

At Design point the coefficient of lift yields induced drag as follows;

$$CL = 1.016$$

$$\mathbf{CDi = 0.03611}$$

Induced drag is also known as drag due to lift and thus with the value of span efficiency e=0.65 having a constant values of aspect ratio the induced drag coefficient becomes the function of the lift coefficient directly and its variation is directly proportional to the variation in lift coefficient as shown below in the plot.

Plot of lift coefficient versus induced drag

Induced drag variation with respect to lift coefficient is shown below.

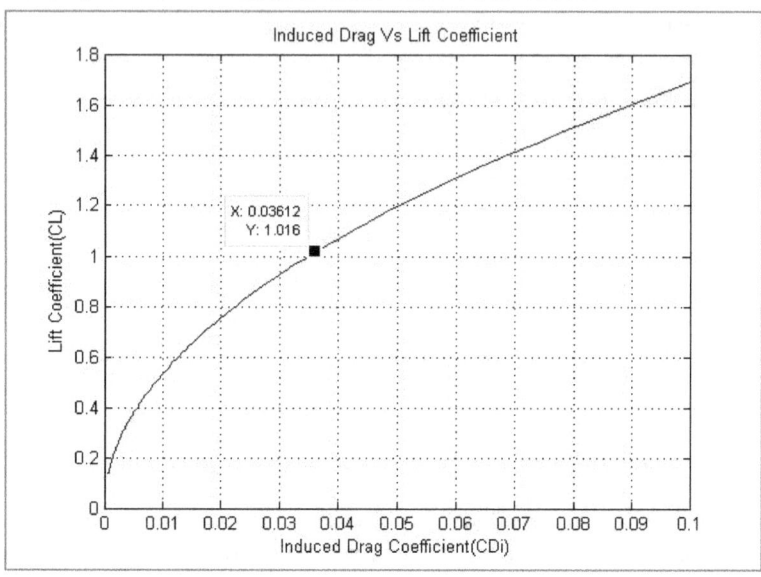

[Fig 9.3: CL vs. CDi plot]

9.2.1.2 Skin Friction drag

This is a drag on a body resulting from viscous shearing stresses over its wetted surface.

Skin friction drag is calculated as;
Laminar flow is considered at the leading edge and the relation used for calculation is;

$$C_{fe} = \frac{1.5 \times 1.328}{(Re)^{0.5}} \text{ (For Laminar Flow)}$$

Skin friction drag comes out to be;

$$C_{fe} = 0.003088$$

Turbulent flow is considered at the trailing edge of the airplane and skin friction is calculated is as follows:

$$C_{fe} = \frac{1.5 \times 0.42}{ln^2(0.056\,Re)} \quad \text{(For Turbulent Flow)}$$

Cfe =0.00622

Total skin friction drag estimated is;

Cfe=0.003088(Laminar)

Cfe=0.00622(Turbulent)

Total Cfe=0.003088+0.00622=0.009308

Cfe=0.009308

Plot of skin friction drag coefficient versus Reynolds number

Skin friction drag variation with respect to variation in Reynolds number at different altitude is shown as below:

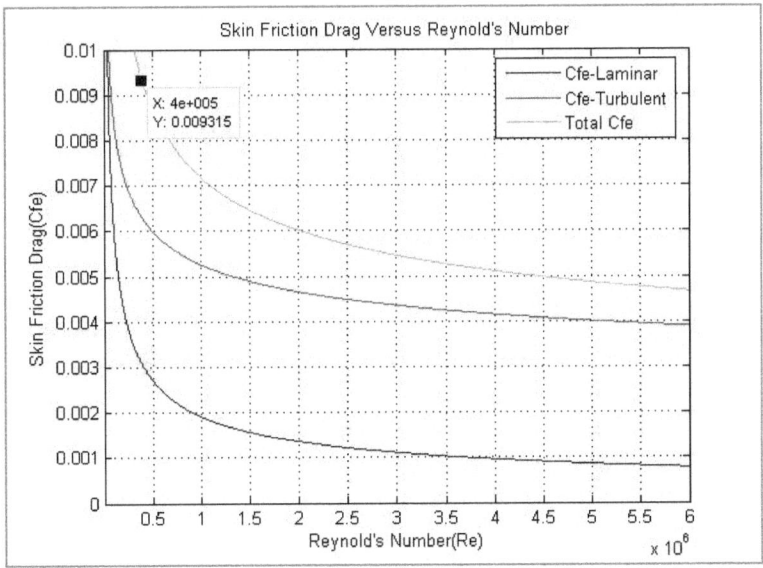

[Fig 9.4: Skin friction drag coefficient versus Reynolds number plot]

9.2.1.3 Parasite drag

It is the total drag of the airplane minus the induced drag thus, it is the drag not directly associated with the production of lift.

Parasite drag coefficient C_M is approximated as:

$$C_M = C_{fe}(S_{wet}/S_{ref})$$

$$(S_{wet}/S_{ref}) = \left(2 + \frac{\frac{t}{c}}{3}\right) = 2.005$$

$$t/c = 0.01516$$

$$Cf_e = 0.009308$$

$$\mathbf{C_M = 0.018616}$$

9.2.1.4 Form drag

The drag on a body resulting from the integrated effect of static pressure acting normal to its surface resolved in the drag direction.

$$C_{FORM} = \frac{\sum\left(C_f \times FF \times Q \times S_{wet}\right)}{S_{ref}}$$

For foam factor calculation we have the following relation;

$$FF = \left[1 + \frac{0.6}{(x/c)_m}\left(\frac{t}{c}\right) + 100\left(\frac{t}{c}\right)^4\right] \times [\,1.34\,M^{0.18}\,(\cos\Lambda_m)^{0.28}\,]$$

t/c =0.01516   *(x/c) m=0.3×Chord=1.84*   *M=0.1*

Thus the form factor comes out to be;

$$FF = 0.88966$$

Q=1, Swet/Sref=2

$$\mathbf{C_{FORM} = 0.01656}$$

9.2.1.5 Interference drag

The increment in drag resulting from bringing two bodies in proximity to each other is the interference drag. For example, the total drag of a wing-fuselage combination would usually be greater than the sum of the wing drag and fuselage drag independent of each other.

In our case there is fuselage thus a symmetric flying wing configuration enables us to neglect the interference drag.

$$C_{INT} = 0$$

9.2.1.6 Trim drag

The increment in drag resulting from the aerodynamic forces required to trim the airplane about its centre of gravity. Usually, this takes the form of added induced and form drag on the horizontal tail.

Absence of tail surfaces and in our particular aircraft there is no such trim drag thus,

$$C_{TRIM} = 0$$

9.2.1.7 Profile drag

It is usually taken as mean the total of the skin friction drag and form drag for a two dimensional airfoil section.

9.2.1.8 Cooling drag

It is the drag resulting from the momentum lost by the air that passes through the power plant installation for purposes of cooling the engine oil and accessories.
No such drag in our case as not an engine powered airplane and electric propulsion system don't have such type of cooling drag.

9.2.1.9 Base drag

The specific contribution to the pressure drag attributed to the blunt after-end of a body.

There is no any blunt after end in the aircraft configuration as a smooth and symmetric wing so base drag is also zero in our case.

$$C_{BASE}=0$$

9.2.1.10 Wave drag

Limited to supersonic flow, this drag is a pressure drag resulting from the static pressure components to either side of shock wave acting on the surface of the body from which the wave is emanating.

But in our case there is a very low speed flight envelope so wave drag is ignored.

$$C_{WAVE} = 0 \text{ (As low subsonic flight profile)}$$

9.2.1.11 Components Drag

Components like landing gears and propeller have drag which is calculated as follows:

Landing Gears Drag

Landing gears placement on the airplane configuration is optional depending upon the conceptually best suited design. The preliminary drag calculation of drag of landing gears is given as follows;

$$C_{Dlanding\ gear} = 0.98 + \pi CL(\frac{Sgear}{S})$$

$$Sgear = 0.1366 ft^2$$
$$CL = 1.016$$

As two pairs of landing gears are placed so;

$$C_{Dlanding\ gear} = 0.98 + 3.14 \times 1.016(\frac{0.1366 \times 4}{528})$$

$$C_{Dlanding\ gear} = 0.9833$$

So, from the above calculation it seems that the drag is too much even more than that of the aircraft so decision was made not to use landing gears in order to attain the best performance so a free landing and takeoff will be made through making launching mechanism in detail design phase.

Propeller Drag

Calculations of drag due to propeller are as follows:

Motor Selected:

- Hacker A60-24s
- Power=1900 watts

Recommended Propellers

Recommended propellers with the selected motor are as:

- 20.5×12 inches
- 20×13 inches

Wing Milling drags Coefficient due to propeller

The drag produced by the propeller which are that of wind milling and stopped propeller drag. The calculation of wind milling drags coefficients is as;

$$C_{Wm-prop} = 33 \times \left(\frac{1}{qs}\right) S_{HP}/V$$

$$S_{HP} = 2.54\ hp$$
$$V=74ft/s$$
$$q=0.60236,\ S=528ft^2$$

$$C_{Wm-prop} = 0.00010793$$

Drag Coefficient due to stopped propeller

The drag due to stopped propeller is given as:

$$C_{Stp-prop} = 0.00125np \times \left(\frac{Dp^2}{S}\right)$$

$$np = 2$$
$$Dp = 20 inches = 1.6ft$$
$$C_{Stp-prop} = 0.00001212$$

Total Propeller Drag

So, overall propeller drag is given as

$$C_{propeller} = C_{Stp-prop} + C_{Wm-prop}$$

$$C_{propeller} = 0.00001212 + 0.00010793 = 0.00012005$$

$$C_{propeller} = \mathbf{0.00012005}$$

9.2.2 Total Zero Lift Drag Coefficient (C_{Do})

Total lift coefficient by summing all drags comes out to be

$$C_{Do} = C_{FORM} + C_M + C_{INT} + C_{BASE} + C_{WAVE} + C_{TRIM} + C_{Propeller}$$

$$C_{Do} = 0.01861 + 0.01656 + 0.00012 = 0.03529$$

$$\mathbf{C_{Do} = 0.03529}$$

9.2.3 Total Drag Coefficient (C_D)

The total drag coefficient thus comes out to be as follows;

$$C_D = C_{D_o} + K C_L^2$$

Final Drag Equation

$$C_D = 0.03529 + 0.03499 \, C_L^2$$

9.3 Aerodynamics Plots

9.3.1 Polar Curve

The plot between coefficient of lift and coefficient of drag is shown below.

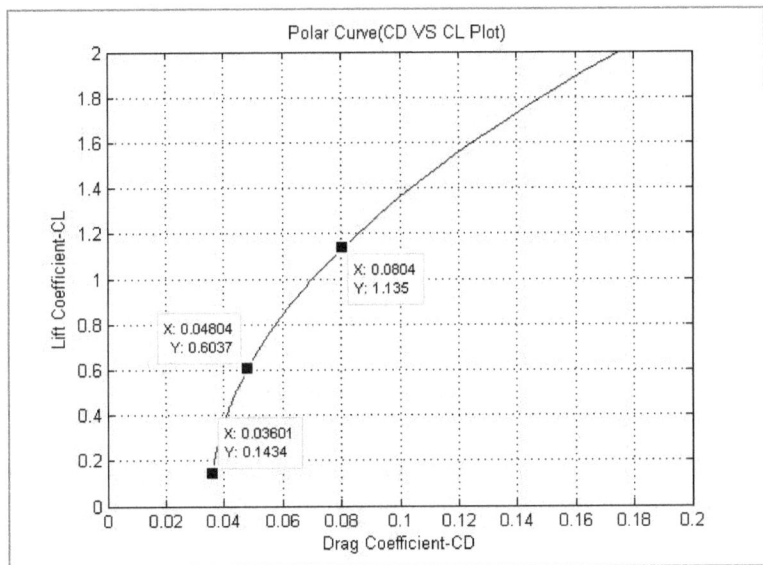

[Fig 9.5: Polar curve]

All aerodynamic plots are plotted at 60,000ft altitude density values.

The plot shown above shows the variation of drag coefficient along the lift coefficient variation where in our design of SAIST the minimum drag coefficient depicted from the plot comes out to be 0.03601 and the maximum drag coefficient comes out to be 0.0804.

9.3.2 Drag vs. Velocity Plot

The plot shown below is the drag variation with respect to the velocity.

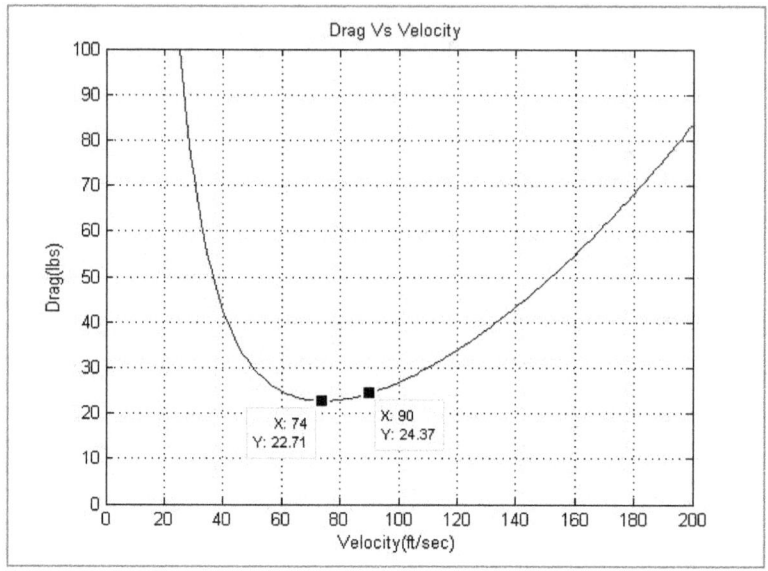

[Fig 9.6: Drag versus Velocity Plot]

So, from plot it is shown that drag at the design speed of 74 ft/sec comes out to be **22.71lbs**.

9.3.3 Lift Coefficient versus Velocity Plot

The plot shown below is the variation curve between lift coefficient and velocity as;

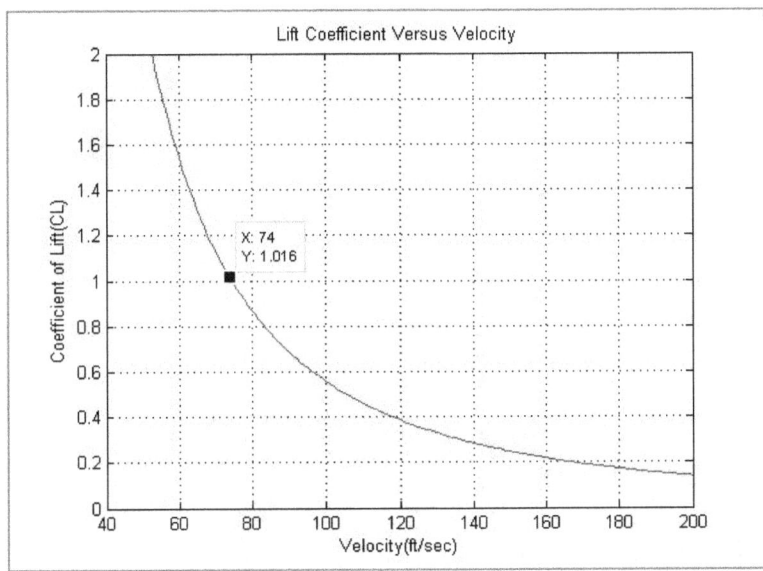

[Fig 9.7: CL versus Velocity Plot]

Thus, coefficient of lift is modified to a value of 1.016 at optimized design speed of 74ft/sec as depicted from the plot shown above.

Chapter 10 | Performance

10.0 Performance

10.1 Constraint Analysis

Constraint analysis is the method to analyze the weight, thrust, and wing area at the same time to get the optimum performance during specific mission conditions. The thrust to weight and wing loading plot were analyzed to obtain the design point. The take-off, landing, stall and cruise constraints are plotted for thrust to weight and wing loading. The following constraints are used to obtain the design point:

- Required altitude and endurance must be attained.
- Lowest energy consumption and maximum energy attainment from solar source.
- Minimum drag and required power at design speeds.
- Take off and landing sequences must be favorable.
- Maximum lift coefficient and aerodynamic efficiency at design speed.
- The structure should withstand the load factor calculated at operating speed regimes
- The aircraft must sustain the 360 degree turn at cruise speed.

10.2 Aerodynamic Efficiency

The most important factor in performance of an aircraft is the lift to drag ratio as it's the overall aerodynamic efficiency of the aircraft and also an idea of increment in lift with reference to drag and the variation of both as a function of dynamic pressure variation and velocity variation.

$$\frac{L}{D} = \frac{1}{(\frac{q C_{D,0}}{\frac{W}{S}}) + (\frac{W}{S})(1/\pi q A e)}$$

$$q = 1/2 \times \rho \times V^2 = 0.60236 lb/ft^2$$

$$e = 0.65 \quad , \quad W/S = 0.612 lb/ft^2$$

$$C_{D,0} = 0.035291$$

$$L/D = 14.226$$

The value of L/D calculated above is that of L/D at cruise conditions where the cruise velocity comes out to be 74 ft/sec as it's the optimized value of speed at which L/D comes out to be of maximum value.

10.2.1 Plot of L/D Vs Velocity

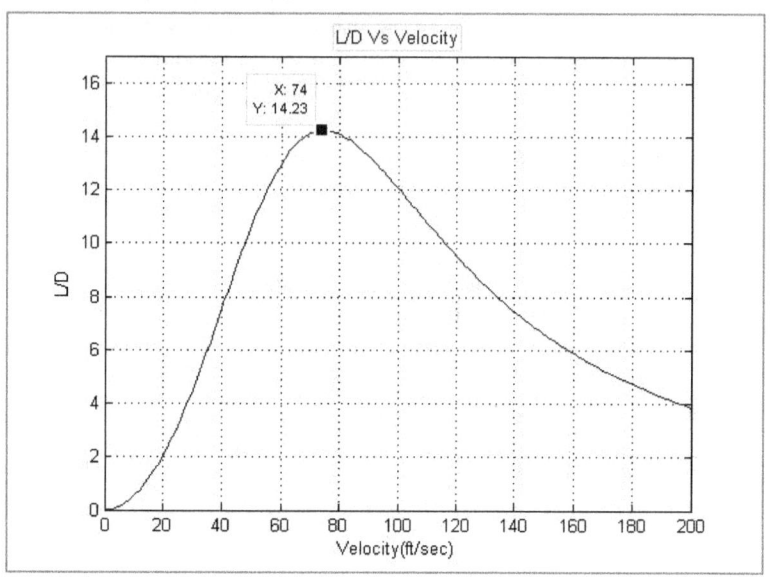

[Fig 10.1: L/D variation plot with velocity]

Thus maximum L/D comes out to be at 74ft/sec which is the design speed for SAIST.

10.2.2 Thrust to Weight ratio Variation with L/D

In level flight thrust to weight ratio:

$$T/W = 1/(L/D)$$

The plot shown below shows the variation between thrust to weight ratio and lift to drag ratio.

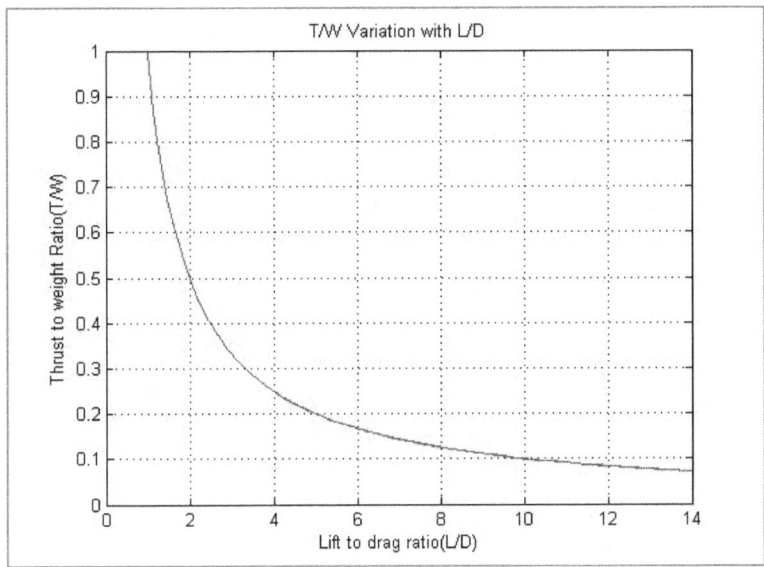

[Fig 10.2: T/W variation with L/D ratio]

10.2.3 Aerodynamics Coefficients Plot

Below a graph is drawn between velocity and two aerodynamic ratios which govern the minimum thrust required and minimum power required values. As can be seen that minimum velocity for minimum thrust required occur at maximum C_L/C_D and minimum velocity for minimum power required occurs at maximum. $C_L^{3/2}/C_D$

$$C_L^{1/2}/C_D \quad , \quad C_L^{3/2}/C_D \quad , \quad C_L/C_D$$

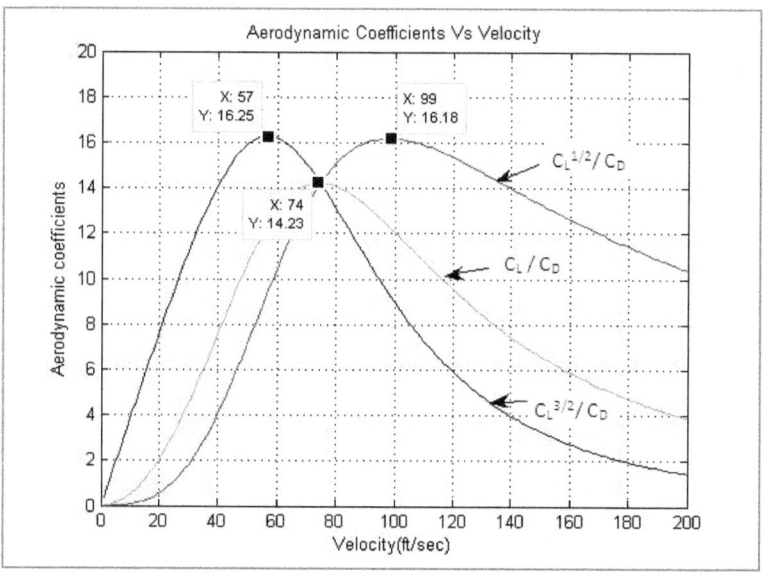

[Fig 10.3: Aerodynamic Coefficients Versus Velocity Plot]

Above graph is drawn between velocity and aerodynamic ratios which govern the minimum thrust required and minimum power required values. As can be seen that minimum velocity for minimum thrust required occur at maximum C_L/C_D and minimum velocity for minimum power required occurs at maximum $C_L^{3/2}/C_D$ .So parameters come out to be

- Velocity (Minimum Thrust) =74ft/sec
- C_L (Minimum thrust) = 1.01
- Velocity (Minimum power) =57ft/sec
- C_L (Minimum power) =0.785

10.3 Thrust required profile Vs Velocity

Net thrust required Vs Velocity

$$T_R = qSC_{Do} + qSKC_L{}^2$$

$$C_L = \frac{W}{1/2\rho_\infty V_\infty^2 S}$$

Thrust required at different velocities is analyzed on velocity versus thrust required plot;

$$T_R = Drag = q_\infty SC_D = \frac{W}{C_L/C_D}$$

$$Parasite\ T_R = q_\infty SC_{DO}$$

$$Induced\ T_R = q_\infty SKC_L^2$$

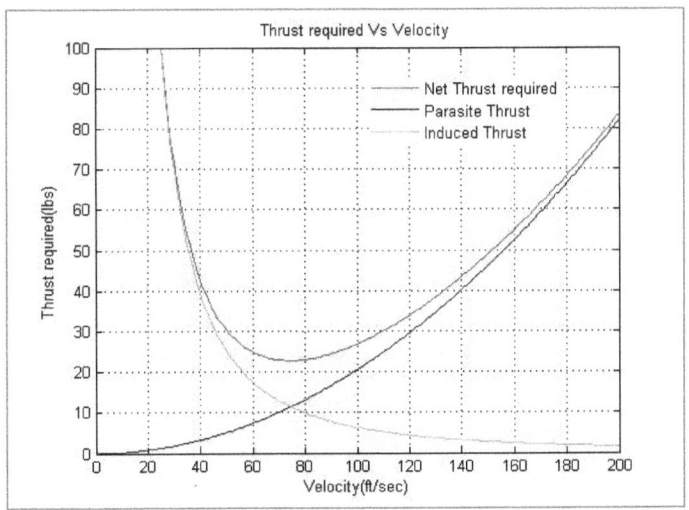

[Fig 10.4: Thrust required versus Velocity Plot]

So, induced and parasite drags are equal at **74ft/sec.**

10.4 Power Required Vs Velocity

The power required for an airplane is a level un accelerated flight at a given altitude (sea level in our case) with a certain velocity is given by:

$$P_R = T_R = \frac{W}{C_L \big/ C_D} V_\infty$$

Power available plot versus velocity in ideal case without considering propeller power losses.

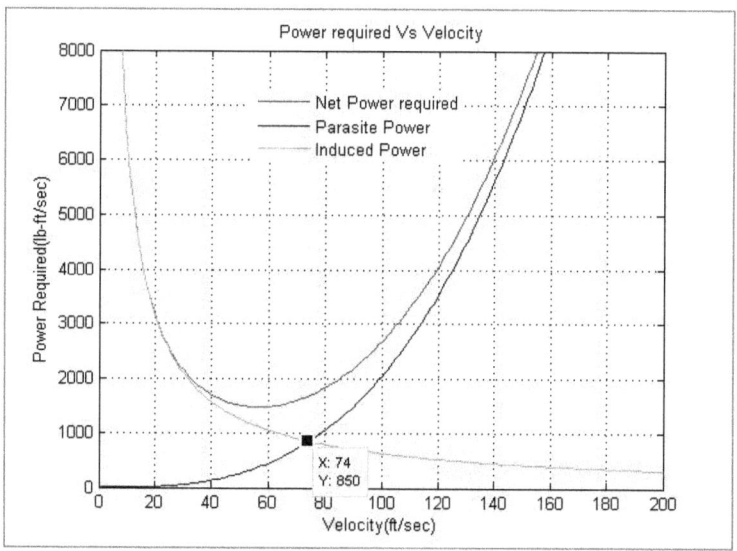

[Fig 10.5: Power required versus Velocity]

Maximum velocity comes out to be **74ft/sec.**

10.4.1 Power Available and Power Required

The variation of the power required versus velocity is shown below. It can be seen from the curve that minimum power required is very close to the design velocity of 74f/s.

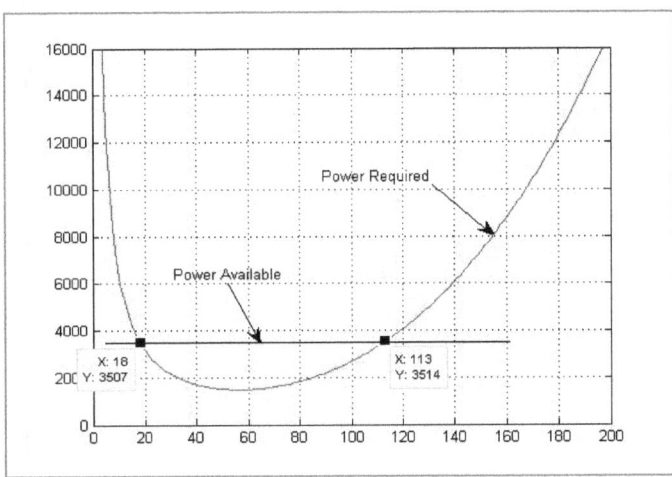

[Fig 10.6: Power versus Velocity Plot]

10.5 Climb Performance

In the overall mission of the aircraft, there is a climb phase in which the aircraft increases its height to the required cruising level and a descend phase from the end of the cruise to the landing. In these phases of the flight, the difference between the propulsive thrust and the airframe drag is used to change the potential energy, and the kinetic energy of the aircraft. If the thrust exceeds drag, the aircraft will climb and if the drag exceeds thrust, it will descend; the rate at which this occurs will depend on the relative magnitudes of the thrust and drag forces. Although climb and descent imply changes in height, they may also involve changes in speed since the air density decreases with the altitude.

10.5.1 Rate of Climb *vs.* Velocity

Climb performance is important from both economic and flight safety points of view. In a climb, the potential energy of the aircraft is increased and fuel energy must be expended to achieve this. The fuel required to climb to a given height can be minimized by the use of correct climb technique and optimum economy of the operation can be attained.

The safety of the aircraft depends on its ability to climb above obstructions at all points on the flight path. Sufficient excess thrust must be available to ensure that the aircraft can meet certain minimum gradients of climb in any of safety critical segments of the flight.

Rate of climb is estimated by the following relationship as:

$$ROC = \frac{ExcessPower}{W} = \frac{P_A - P_R}{W}$$

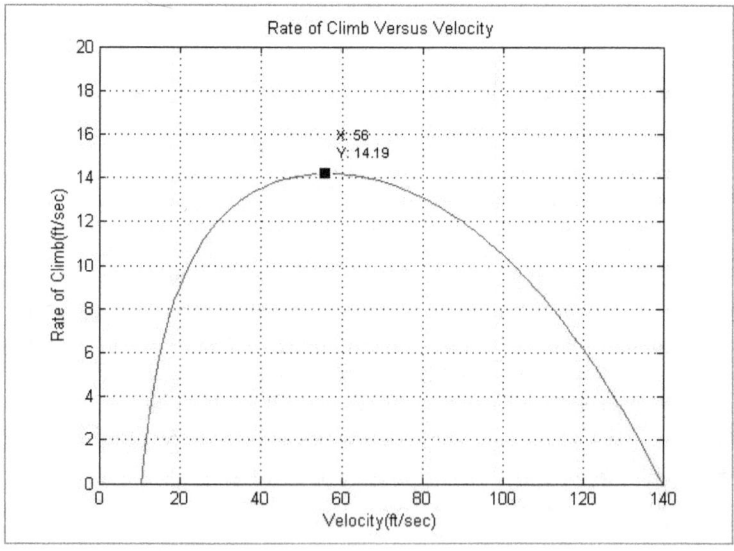

[Fig 10.7: Rate of climb versus Velocity]

So, it is shown in the graph above that maximum rate of climb comes out to be at **56ft/sec**.

10.5.2 Climb Angle

The climb angle is calculated as

$$\sin \gamma_{\text{Climb}} = \frac{T}{W} - \frac{1}{L/_D}$$

$$\sin \gamma_{\text{Climb}} = 0.242 - 1/14.22$$

$$\gamma_{\text{Climb}} = 9.888 deg = 0.1713 radians$$

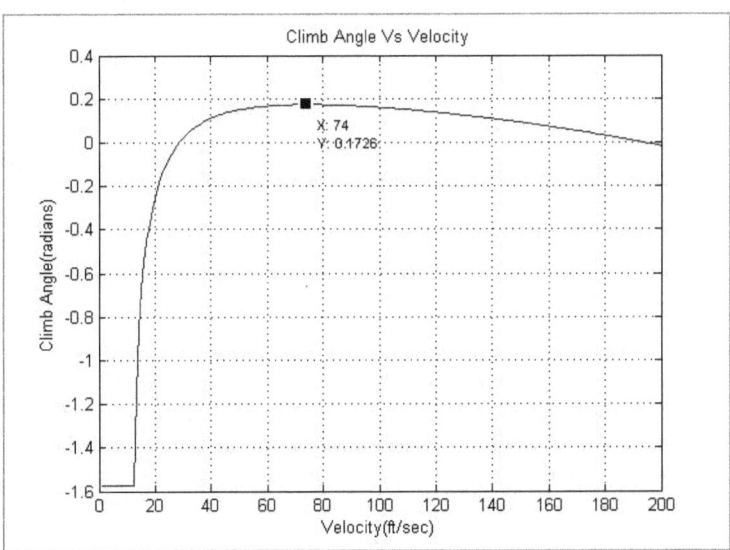

[Fig 10.8: Climb angle versus Velocity]

The maximum climb angle comes out to be at 74ft/sec.

10.6 Gliding Performance

10.6.1 Glide Path Angle

When the propulsive thrust is less than the airframe drag, then the aircraft will decelerate or descend. The descending flight path can be varied from a shallow descent to a very steep descent either by reducing the engine thrust or by increasing the airframe drag. The drag can be increased either by aerodynamic means or by varying the airspeed. In the special case of gliding flight, in which there is no propulsive thrust, the descent will be determined by the lift to drag ratio. In this case, the minimum rate of descent occurs at the minimum power speed and the minimum gradient occurs at the minimum drag speed. The minimum glide angle is obtained at maximum L/D ratio which corresponds to the maximum range for glide.

$$\tan \theta_{min} = \frac{1}{(L/D)_{max}}$$

$$\theta_{min} = \tan^{-1}\left(\frac{1}{17.825}\right)$$

$$\theta_{min} = 3.210° = 0.0556 radians$$

10.7 Take-off Performance Analysis

All conventional aircrafts and UAV flights start at the point of departure with a takeoff and end at the destination with a landing; these are known as the terminal phases of the flight.

In takeoff phase, the aircraft is transferred from its stationary, ground-borne, state into a safe airborne state. Similarly, in the landing phase, the aircraft is transferred back from the airborne state to the ground-borne state and brought to a halt. Since these maneuvers take place in close proximity to the ground, and at a low airspeed, there is a relatively high risk to the safety of the aircraft.

The terminal phases of the flight path consist of two parts. In the first part, the ground-run distance, the aircraft is in contact with the ground and its weight is supported, at least partly by the landing gear. In the second part, the airborne distance, the aircraft is in transition between the ground-borne state and safe airborne flight. Ground handling qualities of the aircraft matter a lot in these phases.

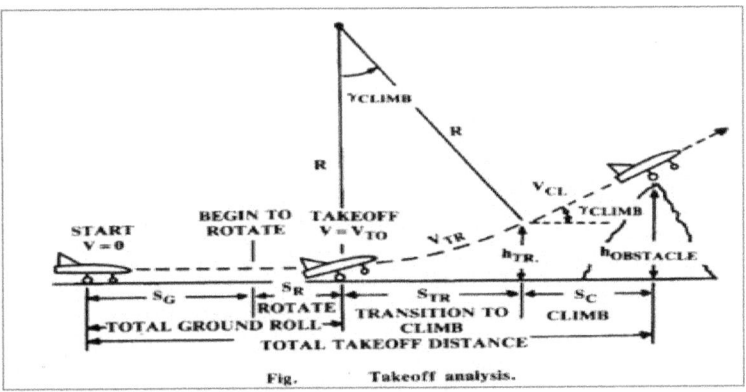

[Fig 10.9: Takeoff Analysis, Raymer]

Takeoff distance comes out to be

$$S_G = \frac{1.21 \text{ x } \dfrac{W}{S}}{g\rho_\infty C_{L,max}\left(\dfrac{T}{W}\right)}$$

$$S_T = \sqrt{R^2 - \left(R - h_{TR}\right)^2}$$

Total take off distance = S_G+S_R+S_{TR}+S_C

S_G =379.744ft

Speed during transition=1.1Vstall=76.9505ft/sec

Stall Velocity=69.955ft/sec

Take off speed=76.9505ft/sec

Climb Speed=1.2Vstall=83.946ft/sec

Average velocity during transition=1.15Vstall=104.9325ft/sec

S_R=V_{TO}=76.95 ft (For small Lighter Airplanes)

R = 0.205 × (V $_{Stall}$)^2= 1003.2089ft

$$\text{Climb Angle} = 0.1713 \text{ radians}$$

$$S_T = R\sin\gamma_{climb} = 172ft \qquad \text{or}$$

$$h_{TR} = R\,(1\text{-}\cos\gamma_{climb}) = 1.86564ft$$

Or other way results transition distance as

$$S_T = (R^2 - (R - h_{TR})^2)^{1/2} = 172.3248ft$$

As obstacle is not cleared during transition so climb distance is estimated as

$$S_C = h_{\,Obstacle} - h_{TR}/\tan\gamma_{climb} = 201.3497ft$$

So takeoff distance is

$$SG + SR + STR + SC = 379.744 + 76.95 + 172.324 + 201.349 = 830.367ft$$

Total Take off Ground Distance=830.367ft

10.7.1 Accelerated Climb

The energy required to climb to altitude and accelerate to flight velocity was the sum of the change in kinetic and potential energy and the energy dissipated by drag divided by an average efficiency of the power plant between lift off and cruise velocity. The drag during accelerated climb was approximated with the induced drag evaluated at the average lift coefficients of the start and finish velocities.

The ground distance of the accelerated climb (Sac) was calculated using an effective net force (Fac) during transition which was an average of the thrust available and thrust required at the lift off (V_{LO}) and cruise velocities (V_c).The relation from reference is as follows;

$$Sac = W/F_{ac}(\,h + Vc^2 - V_{LO}^{\,2}\,)/2g$$

$$Fac = 0.5\times\,(T_{ALO}\text{-}T_{RLO} + TAC\text{-}T_{RC})$$

$$T_{ALO} = 31.2162$$

$$T_{AC} = T_{ALO}$$

$$T_{RLO} = 23.684lb$$

$$T_{RC} = 22.7083lb$$

$$h=1000ft \text{ (Initial Climb altitude)}$$

So, substituting values yield average net force which is given as

$$Fac=16.0401lbs$$

So ground distance of accelerated climb comes out to be

$$Sac = 173.511ft$$

10.7.2 Landing performance

In the landing phase, the aircraft is transferred back from the airborne state to the ground-borne state and brought to a halt. The landing distance profile is as:

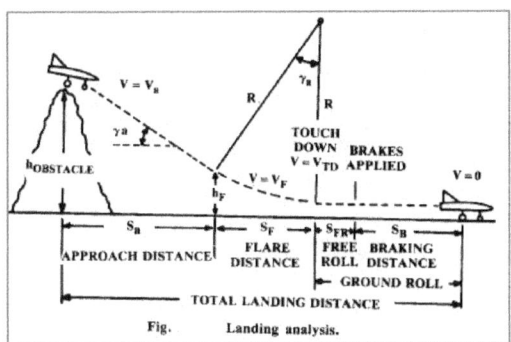

[Fig 10.10: Landing Analysis, Raymer]

The landing distance is given as

$$\text{Landing Distance} = S_A + S_F + S_{FR} + S_B$$

$$\text{Approach Speed} = Va = 1.3Vstall = 90.9415ft/sec$$

$$\text{Touchdown speed} = 322.944ft/sec$$

$$\text{Flare Velocity} = 1.15Vstall = 80.448ft/sec$$

$$\text{Approach Angle} = 9.88degrees = 0.71radians$$

R=1003.2089ft (R comes out to be during landing)

$$Sa = 50-h_f/\tan\theta=201.349ft$$

$$S_f = 172.324ft$$

$$S_{fr} = 76.95ft$$

$$S_b = 349.368ft$$

So, Landing Distance comes out to be;

Landing Distance $= 201.349 + 172.324 + 76.95 + 349.368 = $ **800ft**

Same as that of take off distance as weight doesn't change at the end of mission profile.

10.8 Turning Performance

The most important parameters in turning performance are:

- Turn Rate &
- Turning Radius

10.8.1 Load Factor

The overall performance of the aircraft is limited by the structural strength of the airframe. The basic reasons for this are

- The pressure loading produced by the dynamic pressure of the airflow increases with the square of airspeed and with it the air loads on the structural components of the aircraft.

- The normal acceleration associated with the maneuvering flight produces structural loads in the airframe. The maximum allowable load factor is determined by the load bearing capability of the airframe structure.

For better performance in case of fighter aircraft that must have the capability of superior dog fight there must be minimum turn radius and maximum turn rate.

$$n = \frac{L}{W} = \left(\frac{L}{D}\right)\left(\frac{T}{W}\right) = 14.22 \times 0.242 = 3.441$$

10.8.2 Maximum load factor

For maximum load factor we have;

$n_{Max} = 3.441$ (*As electric powered so maximum will be at deisgn point*)

10.8.3 Roll Angle

The roll angle is calculated from the relation shown below.

$$\Phi = Arccos\left(\frac{1}{n}\right) = 72.895° = 1.263 radians$$

The variation of the roll angle with velocity is shown in the figure below. It can be seen that the maximum roll angle comes out to be at the design point velocity which is 74 ft/s.

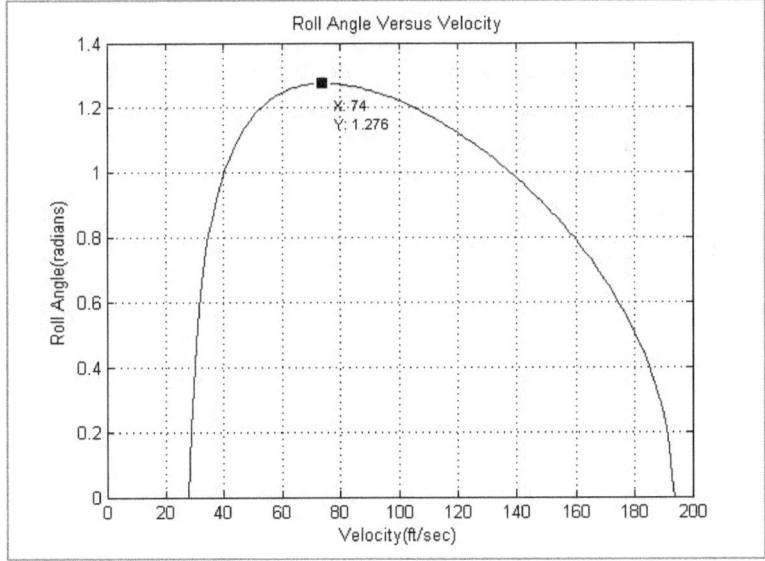

[Fig 10.11: Roll angle versus Velocity]

10.8.4 Maximum Roll Angle

Maximum roll angle is calculated as

$$\Phi = Arccos(1/n_{Max}) = 72.895°$$

10.8.5 Turn Radius

As explained before that for a better performance we need lowest turn radius and highest turn rate thus for lowest turn radius we have to evaluate:

- Highest possible load factor
- Lowest possible Velocity

So, using above mentioned values we can calculate the turn radius as;

$$R = v^2 / g \, (n^2-1)^{1/2} = 51.635ft = 15.73m$$

10.8.6 Turn Rate

For obtaining largest possible turn rate, there are two factors to be considered.

- Highest possible load factor
- Lowest possible Velocity

This is the same criteria as for minimum turn radius. So using relation of turn rate in terms of velocity and load factor as;

$$\omega = g \, (n^2-1)^{1/2} / v = 1.43 radians/sec = 82.511 degrees/sec$$

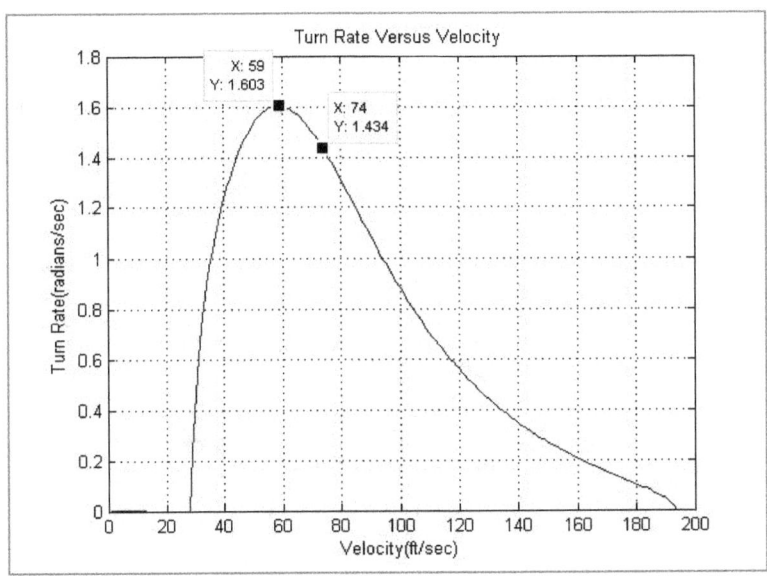

[Fig 10.12: Turn rate versus velocity]

10.8.7 Turning Distance

The distance the aircraft flies during a 360 degree turn is calculated as;

$$St = 2\pi(Vc^2)/\ gTan\ (\emptyset)$$

$$\emptyset = 72.895 degrees$$

$$Vc = 74 ft/sec$$

$$St = 328.65 ft$$

10.9 Pull Up and pull down maneuvers

10.9.1 Pull up maneuver

Considering that in a steady level flight, aircraft experiences a sudden change of lift. As a result it begins to lift upward. For this maneuver, the flight path becomes curved in the vertical plane, with a turn rate. Radius of turn and turn rate can be found as:

$$\text{Pull- Up} \Rightarrow \quad \omega = \frac{g(n-1)}{V_\infty} = 1.0617 \text{rad/sec}$$

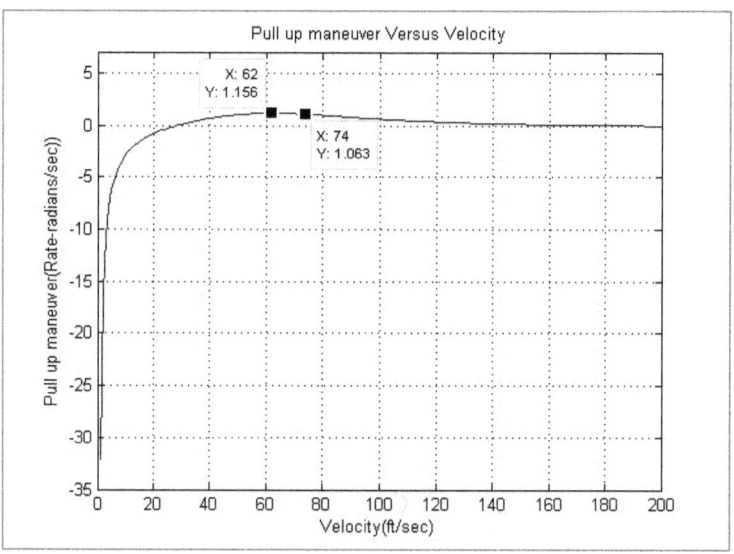

[Fig 10.13: Pull up maneuver plot]

10.9.2 Pull down maneuver

Similarly for pull down maneuver the turn radius and turning rate is as:

$$\text{Pull Down} \Rightarrow \quad \omega = \frac{g(n+1)}{V_\infty} = 1.932 \text{rad/sec}$$

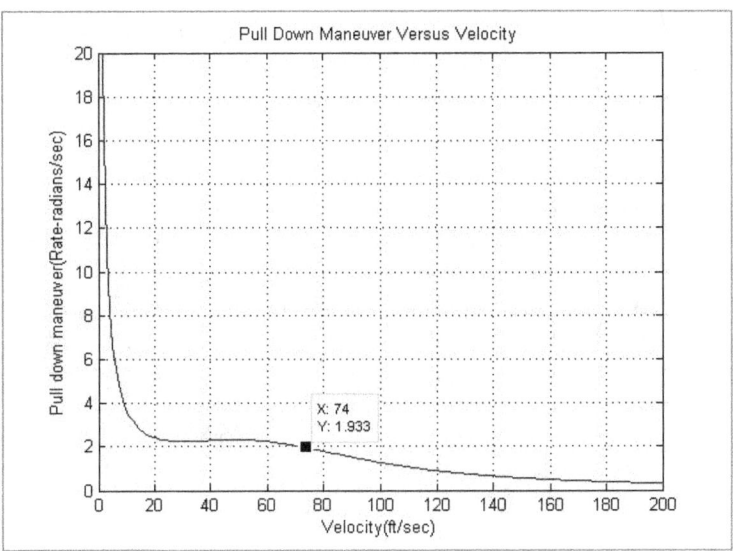

[Fig 10.14: Pull down maneuver plot]

10.10 Range Calculations

We have a propeller driven airplane in which there is no such fuel consumption that is the aircraft is solely dependent upon the batteries and power supply i.e. Solar energy obtained from the panels and useful energy for running motors during flight.

So, range will be a function of velocity and timing of batteries after reaching a certain altitude.

10.11 Propeller performance

Electric powered propeller driven aircraft are greatly influenced by the propeller performance so the performance of propeller selected in our design has been analyzed.

Propeller converts the turning power of propulsion system into the thrust force. This thrust is equal to the mass of air forced backward by the propeller per unit time multiplied by the added velocity imparted to this air.

10.11.1 Slipstream

The slipstream is produced by a propeller producing thrust by forcing air backwards. It is a cylindrical core of spiraling air that flows back over the fuselage and wings. The fact that it strikes the wings has important effects and some of them are detrimental and some are beneficial. The slipstream flow is faster than the free-stream flow; this means that the drag of the fuselage, wings, and other parts exposed to it is larger. The slipstream moves over the wings and is beneficial in providing for effective control by the tail surfaces since the aerodynamic forces produced by these surfaces are dependent on the square of the velocity of the air moving over them. This is important in the cases of taxiing or takeoff when the free-stream velocities may be low.

The rotary motion of the slipstream, however, causes the air to strike the tail plane at an angle and not directly head on. This may have an effect on the stability and control of the airplane. The effects of the rotary motion of the propeller may be counteracted by using contra rotating propellers (spinning in opposite directions).

In our design as there is no tail plane and is solely a flying wing configuration thus slipstream has no very critical effects on the drag and stability of the aircraft.

Efficiency of the propeller is however, influenced by the slipstream phenomena which is approximated in the following calculations.

10.11.2 Propeller Specification

$$\text{Power} = 2.54\text{hp(Single Motor)}$$

$$\text{RPM} = 5772 = 96.2\text{rev/sec}$$

$$\text{Propeller Diameter} = 34 \text{ inches} = 2.83 \text{ ft}$$

Number of blades=2

The design of a propeller, like an airplane, is influenced by many factors, some of which cause contradictions in design. The overall shape is determined by compromise and is largely dependent on the mission to be performed. For low speeds as in our particular design the propeller blade is usually slender with rounded tips. For high speeds, larger paddle-shaped blades are used or more propeller blades are used.

An airfoil can be characterized by relations between *angle of attack*, *lift coefficient* and *drag coefficient*; a **propeller** can be described in terms of **advance ratio**, **thrust coefficient**, and **power coefficient**. The efficiency, which corresponds to the L/D ratio of a wing, can be calculated from these three coefficients. These coefficients are helpful for the comparison of propellers of differing diameters, tested under different operating conditions.

For a wing, properties of propeller are as follows:

10.11.2.1 Advance Ratio
Advance ratio for our aircraft at the design speed is given as

$$J = \frac{V}{nD} = \frac{74}{(96.2 \times 1.75)} = 0.43956$$

Advance ratio is the 'Slip function' and the 'progression factor'.

10.11.2.2 Power Coefficient

The power coefficient in case of propeller is given as;

$$C_P = P/\rho n^3 D^5$$

$$C_P = 550 \times 7.275/ (0.00022 \times (96.2)^3 \times (1.75)^5$$

$$C_P = 1.2446$$

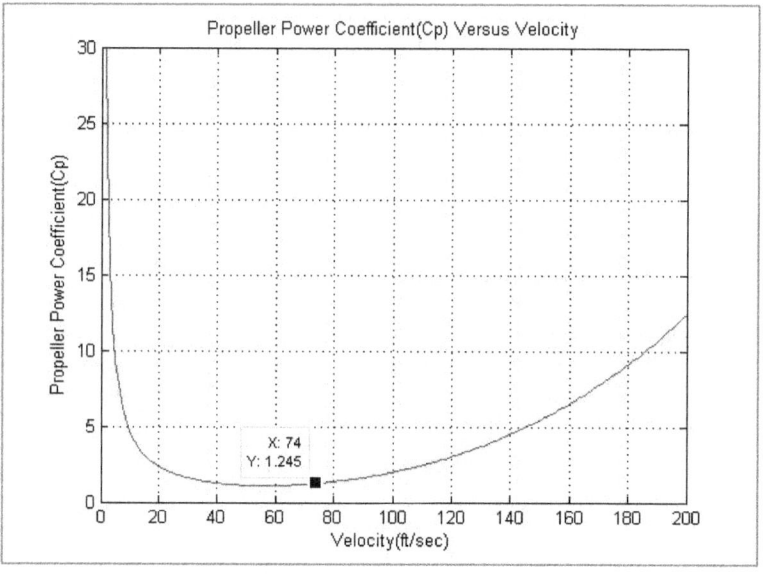

[Fig 10.15: Power coefficient versus velocity plot]

10.11.2.3 Thrust Coefficient

Thrust coefficient of propeller is;

$$C_T = T/\rho n^2 D^4$$

$$C_T = 22.711/0.00022 \times (96.2)^2 \times (1.75)^4$$

$$C_T = 1.1893$$

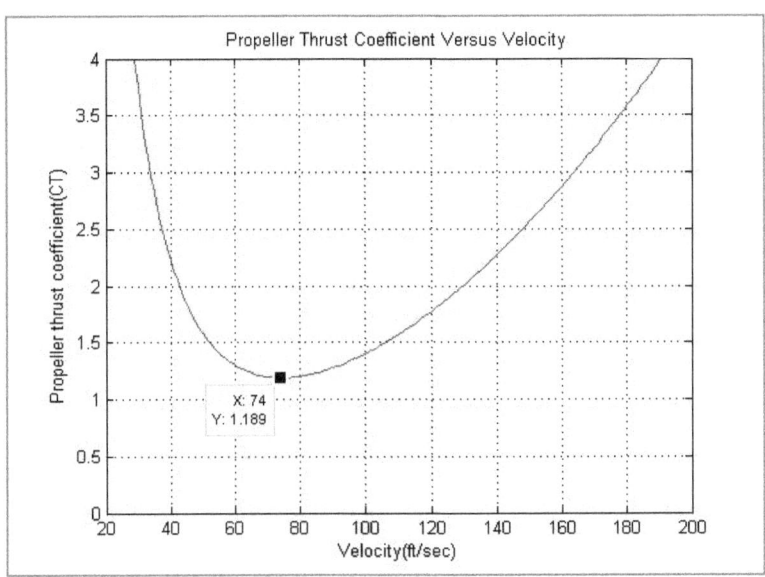

[Fig 10.16: Thrust coefficient versus velocity]

10.11.3 Speed Power Coefficient

The speed -power coefficient is the advance ratio raised to the fifth power divided by the power coefficient. The speed- power coefficient is non dimensional and doesn't involve the propeller diameter, which is useful for comparison between propellers of different sizes.

$$C_{SP} = V(\rho/Pn^2)^{1/5}$$

$$C_{SP} = 0.4207$$

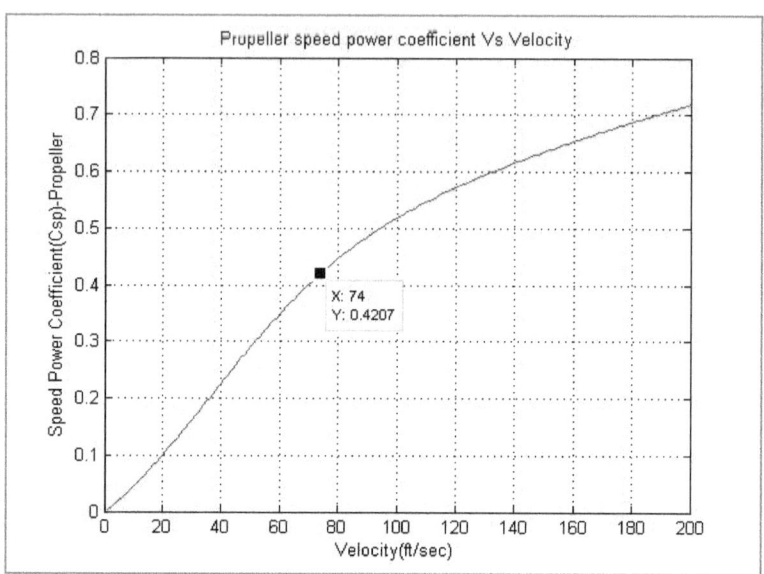

[Fig 10.17: Speed power coefficient versus velocity]

10.11.4 Helix Angle of Flow

The propeller blade consists of a set of airfoil-shaped sections that may vary in outline from the tip to the root of the blade. Although a wing is fixed with respect to the airplane and sees only the relative free-stream flow of air, the propeller is also rotating with respect to the airplane, and it sees an oncoming flow of air which is the vector sum of the airplane free-stream velocity and the propeller rotational velocity. The angle between this relative velocity and the plane of the propeller rotation is called the **helix angle or angle of advance.**

$$Tan\emptyset = \frac{2\pi rn}{V} = \frac{\pi dn}{V}$$

$$Tan\emptyset = 3.14 \times 96.2 \times 1.75/74$$

$$Tan\emptyset = 7.1435$$

$$\emptyset = 81.605 degrees = 1.414 radians$$

For a particular airplane velocity, this helix angle varies from the root to the tip since the tip sections of the propeller are revolving faster than the root sections. As one approaches the root, the helix angle approaches 90°.

10.11.5 Propeller Tip Speed

Maximum speed of the propeller is at the tip which is as;

$$V_t = \pi nd = 528.619 ft/sec$$

$$M_t = 0.4853$$

10.11.6 Propeller Efficiency

Propeller efficiency is a function of power coefficient, thrust coefficient and advance ratio as:

$$\eta = \frac{V}{nD} C_T/C_P$$

$$\eta = 0.42$$

10.11.7 Reduction in Efficiency due to Slipstream

In the slipstream behind a propeller there is a considerable amount of kinetic energy, which was imparted to it by the propulsion system without producing any corresponding propeller thrust. The increased absorption of power reduces the propeller efficiency.

Reduction in efficiency of propeller due to slipstream is calculated as through the following relation;

$$\varepsilon = \sqrt{\frac{1 + C_T - 1}{1 + C_T + 1}}$$

At design point C_T =**1.1893** so, reduction in propeller efficiency is **0.19** i.e. **19%**.

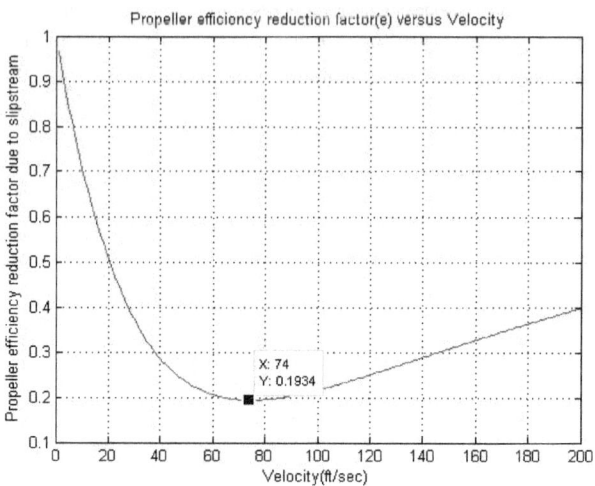

[Fig 10.18: Propeller efficiency reduction factor versus velocity]

V-n Diagram:

The V-n diagram of SAIST is as;

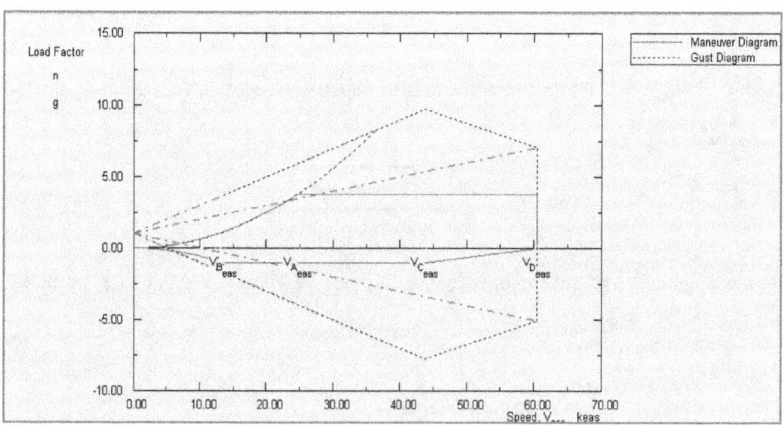

[Fig 10.19: V-n Diagram]

10.12 Energy Curves

Energy curves for SAIST aircraft are as follows:

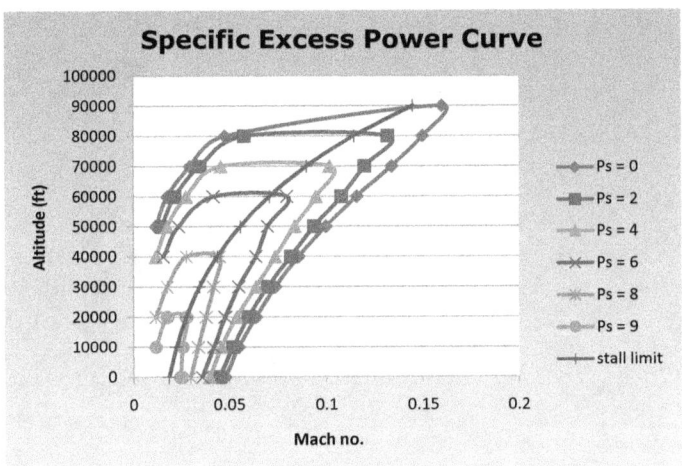

[Fig 10.20: PS Curves]

The specific excess power curves are showing the variation of a constant value of surplus energy (P_s) with the varying altitudes and Mach number. It can be seen that the curve for P_s=0 has its highest value of altitude of 90,000ft which means theoretically, the aircraft is capable of flying up to 90,000fts at which its surplus power becomes zero. Furthermore, the line intersecting the surplus energy curves is the stall limit line at various altitudes. This line and the end portion of the line for P_s=0 form a flight envelope. The lower the aircraft in this envelope, the more it has got surplus energy and the more it can climb and perform maneuvers. With the information of Mach number and altitude, we can determine the excess power that the aircraft has to climb further.

Chapter 11 | Stability

11.0 Stability

To ensure smooth flight of the flying vehicle, it is necessary that the flying vehicle should have enough stability so that any disturbance it may face during its mission profile; the flying vehicle should not deviate from its required path rather it should stabilize itself and return to its original position as it was before the disturbance. There are two main origins for stabilizing the aircraft: first, the aircraft should be structurally designed in such a way that it is *inherently stable* i.e. there is no major role of the control system to stabilize the aircraft again. Secondly, if the aircraft is not inherently stable i.e. It cannot entirely stabilize itself on its own or partially stabilize it then there is a requirement of the *control system* to stabilize the aircraft. The amount of stability required is defined by the aircraft mission.

A flying wing is theoretically the most efficient aircraft configuration from the aerodynamics and structural weight point of view. The benefits are provided by the absence of the major aircraft components except the wing. Less number of components means less drag which leads to the less power requirement of the aircraft for a specific mission. Apart from the absence of the components and benefits the configuration provides, in practice the wing must provide the necessary stability and control requirements of the aircraft that are dictated by its mission which imposes additional constraints on its design. Therefore, these constraints may partially or completely negate the drag and weight benefits in order to provide the necessary stability and control requirements. A flying wing configuration suffers more stability problems as compared to conventional configurations.

A successful flying wing results from a careful balancing of its center of gravity, pitching moment and combination of sweep angle and wing twist.

11.1 Center of Gravity (C.G)

Stability is dependent upon the location of the center of gravity- the more forward the center of gravity is, the more stable the flying wing would be. An important factor defines the stability of an aircraft called the stability margin which denotes the distance between the center of gravity and the aerodynamic center. The aerodynamic center should lie at 25% of the mean aerodynamic chord. If the center of gravity lies at the aerodynamic center then the aircraft would have a zero static margin and if faced a stall situation, then it would not recover itself rather it would descend like a parachute. Therefore, a proper determination of the location of center of gravity plays a vital role in the stability of the aircraft. The more forward the C.G is, the more stable the aircraft would be but there is a limitation for moving the center of gravity more ahead of the aerodynamic center as the aircraft would be pushed down and more force would be required to pitch it up and more twist will be required.

11.2 Pitching moment

As the center of gravity is required to be ahead of the aerodynamic center, a nose down force is generated which has to be counteracted by a force. A reflex chambered airfoil has a positive pitching moment and this moment should counteract the moment produced as a result of the forward location of the C.G. A conventional airfoil has a strong negative pitching moment which would rather assist the nose down moment of the wing causing instability. Therefore, a reflex chambered airfoil should be used to get the required stability. If the moment generated by the reflex chambered airfoil is not sufficient to counteract the nose down pitching moment, then some additional control like flaps or elevators are required to generate the required moment.

11.3 Combination of sweep angle and twist

Using a conventional airfoil requires a pitch up moment which then requires twist to provide the necessary lift for pitching up the aircraft. If the twist is larger, then it can cause some structural problems as well as bringing a portion of the wing more towards stalling angle of attack. The twist required can be reduced by giving an appropriate sweep angle to the wing. By giving the sweep, the wing tip moves away from the aerodynamic center which then provides a larger moment arm. Therefore a proper combination of sweep angle and twist is determined to ensure the stability.

11.4 Role of airfoil

The reflex chambered airfoil provides the longitudinal stability to the flying wing. To understand this, two things are important:

11.4.1 Total force and moment

The pressure forces acting on the surface of the wing section can be replaced by a single force and single moment; both of these acting at a single point that is aerodynamic center which normally lies at the quarter chord, c/4 point of the airfoil. When the angle of attack of the airfoil changes, the moment remains constant and the force changes i.e. the force increases as the angle of attack is increased.

11.4.2 Center of gravity

The 'free floating' bodies perform translations and rotations about their center of gravity. As the aircraft changes its angle of attack, the airplane rotates about its center of gravity.

The following table compares a conventional, chambered airfoil with a reflex chambered airfoil in equilibrium and disturbed state.

Equilibrium state	
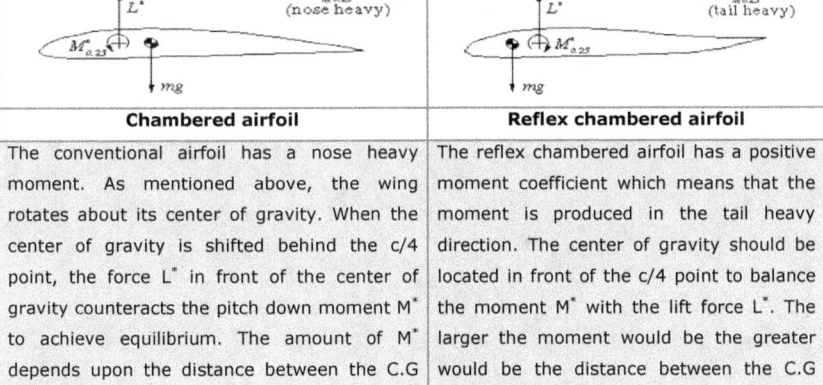	
Chambered airfoil	Reflex chambered airfoil
The conventional airfoil has a nose heavy moment. As mentioned above, the wing rotates about its center of gravity. When the center of gravity is shifted behind the c/4 point, the force L^* in front of the center of gravity counteracts the pitch down moment M^* to achieve equilibrium. The amount of M^* depends upon the distance between the C.G and c/4 point.	The reflex chambered airfoil has a positive moment coefficient which means that the moment is produced in the tail heavy direction. The center of gravity should be located in front of the c/4 point to balance the moment M^* with the lift force L^*. The larger the moment would be the greater would be the distance between the C.G and the c/4 point.

Disturbed state	
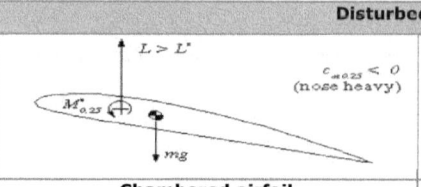	
Chambered airfoil	**Reflex Chambered airfoil**
When a disturbance changes (increases) the angle of attack of the airfoil, the lift force increases and becomes greater than the lift force of the equilibrium state i.e. $L>L^*$ and the tail heavy moment becomes larger than the moment about the c/4 point which is same i.e. $M=M^*$. The wing will pitch up and increase the angle of attack further. The behavior is unstable and a tail is required to stabilize the system.	When the angle of attack increases due to some disturbance, the lift force acting behind the C.G increases i.e. $L>L^*$. As the C.G is ahead of the c/4 point and the rotation is produced about the C.G, the increased lift force will pitch down the airfoil reducing its angle of attack, until the equilibrium state is reached again which shows that the system is stable.

[Table 11.1: Airfoil Stability Comparison]

Keeping in view all the above discussion, necessary steps were taken to structurally design the aircraft in such a way that it is inherently stable and respond to the disturbances smoothly.

A reflex chambered airfoil *mhmi2* is chosen to provide the flying wing with adequate stability without a tail. The moment curve of the airfoil is shown in the airfoil section and is shown here again for convenience.

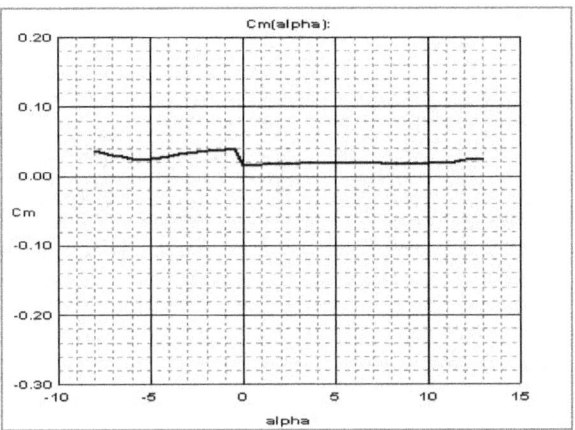

[Fig 11.1: Moment coefficient Vs angle of attack curve]

From the moment curve, it can be seen that the moment coefficient at zero angle of attack Cm_o is positive and is equal to 0.0151. Furthermore, the straight (plank) wing with a reflex chambered airfoil configuration, as we are using in the aircraft, is able to stabilize itself.

11.5 Stability Margin

As defined earlier, stability margin is the distance between the c.g and aerodynamic center of the aircraft divided by the chord of the wing. Mathematically,

$$\text{Stability Margin} = \text{S.M} = \frac{X_{cg} - X_{ac}}{c} \%$$

A small stability margin will produce a small straightening moment as a result of the disturbance. If the c.g is close to the n.p, the straightening moment is small and the wing returns (too) slowly into its equilibrium condition. If the distance Xc.g -Xn.p is large, the c.g is far ahead of the c/4 point and the wing returns quickly to the equilibrium angle. We will require larger flap deflections to control the model, though. If the distance is too large, the wing may become over stabilized, overshooting its trimmed flight attitude and oscillating more and more until the plane crashes. Therefore, a reasonable value of stability margin is chosen. For a slow moving highly stable flying wing, the S.M range is between 5% and 8%. For initial analysis, a S.M of 7% was selected and the results were analyzed using Advance Aircraft Analysis 2.3 (AAA).

11.6 Aerodynamic Center

Aerodynamic center is a point on the airfoil at which the moment remains constant regardless of the change in the angle of attack. For a high-aspect-ratio wing, the subsonic aerodynamic center will be located at the percent mean aerodynamic chord (MAC) of the airfoil aerodynamic center. For most airfoils, it is located at the quarter-chord point (±1%).

The Aerodynamic center was calculated using the AAA software by putting the required inputs. X_{ac} came out to be,

$$X_{ac} = 1.6ft$$

$$\text{Or } \bar{X}_{ac} = X_{ac}/c = 0.2607$$

$$\bar{X}_{ac} = 26.07\% \text{ of the chord}$$

11.7 Center of Gravity

Knowing the stability margin and the aerodynamic center, the center of gravity of the flying wing can be calculated as;

$$S.M = \frac{X_{ac} - X_{cg}}{c}$$

$$\bar{X}_{Cg} = \bar{X}_{ac} - S.M$$

$$\bar{X}_{Cg} = 0.190 = 19\% \text{ of the chord}$$

11.8 Wing pitching moment

The pitching moment of the wing is largely dependent on the pitching moment of the airfoil which also determines its stability. The following equation is an adjustment for the wing aspect ratio and sweep for a straight wing or an untwisted swept wing at low subsonic speeds.

$$Cm_w = Cm_{0airfoil} \left(\frac{A\cos^2 \Lambda}{A + 2\cos \Lambda} \right)$$

$$A = \text{aspect ratio} = 14$$

$$\Lambda = \text{sweep angle} = 0$$

$$C_{m_0} = 0.0151$$

$$C_{m_w} = \mathbf{0.0132}$$

A twist in the wing adds an increment of approximately -0.001 times the twist angle in degrees for a typical swept wing.

11.9 Downwash and Upwash

The air interacting with the flying aircraft experiences a pressure difference when it comes in the vicinity of the wing. Due to this pressure difference, the air is pulled upwards by the pressure about the wing. This upwash creates a push on the body of the aircraft and turns the flow before reaching the propeller located ahead of the wing. This creates a flow behind the wing in the downward direction with an angle equal to the wing angle of attack but this angle diminishes to a half value when it reaches the aft tail. But in this project, there is no aft tail therefore there would be no significant effects of the downwash in our aircraft. This downwash also varies along the span of the wing with the lowest value at the wing tips. The downwash is also affected by the prop wash.

[Fig 11.2: Flow field around a wing]

This upwash and downwash combination affects the moment coefficient therefore they need to be determined. The upwash derivative with respect to the angle of attack can be estimated from the following figure:

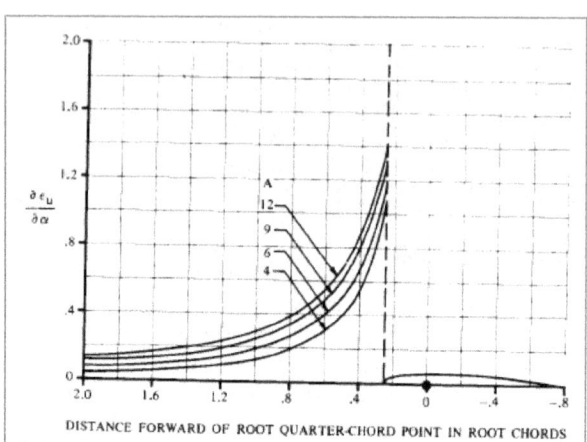

[Fig 11.3: Upwash estimation]

Although the graph does not show the estimation for the aspect ratio of 14 but keeping in view the trend of the lines shown in the graph, the derivative of upwash (ε_u) with respect to angle of attack is estimated as:

$$\frac{\partial \varepsilon_u}{\partial \alpha} = 0.95$$

The downwash angle (ε) derivative is determined from the following figure;

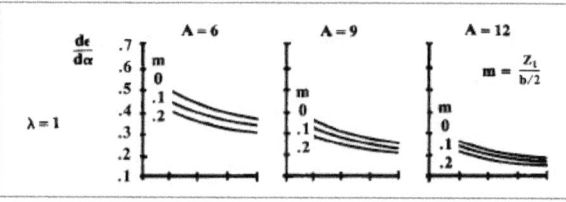

[Fig 11.4: Downwash estimation (M=0)]

The downwash angle (ε) is estimated as

$$\frac{\partial \in}{\partial \propto} = 0.25$$

Considering the effect of downwash and upwash, the resulting angle of attack can be determined by adding an upwash or subtracting a downwash from the free stream angle of attack. The angle of attack derivatives are therefore expressed as;

$$\text{Upwash: } \frac{\partial \alpha_u}{\partial \propto} = 1 + \frac{\partial \in_u}{\partial \propto}$$

$$= 1 + 0.95$$

$$\frac{\partial \alpha_u}{\partial \propto} = \mathbf{1.95}$$

As there is no horizontal tail in our aircraft, the downwash will not have any effect on the stability of the aircraft. Therefore, the angle of attack derivative for the horizontal tail is not calculated here.

11.10 Stability Derivatives

The stability derivatives for the flying wing were calculated using AAA software. The derivatives are distributed into longitudinal and directional stability derivatives.

11.10.1 Longitudinal stability derivatives

- Derivative related to forward speed

C_{L_U}	0.0232
C_{m_U}	0

- Derivatives related to angle of attack

C_{D_α}	0.0008
C_{L_α}	5.3413
C_{m_α}	-0.3741

- Derivatives related to pitch rate

C_{D_q}	0
C_{L_q}	0.0395
C_{m_q}	-0.0101

11.10.2 Directional stability derivatives

- Derivatives related to sideslip

C_{Y_β}	-0.0114
C_{l_β}	-0.0674
C_{n_β}	0.000018

- Derivatives related to roll rate

C_{Y_p}	-0.0012
C_{l_p}	-0.0200
C_{n_p}	-0.1749

11.11 Trim Plot

The plot between moment coefficient and lift coefficient is shown as below:

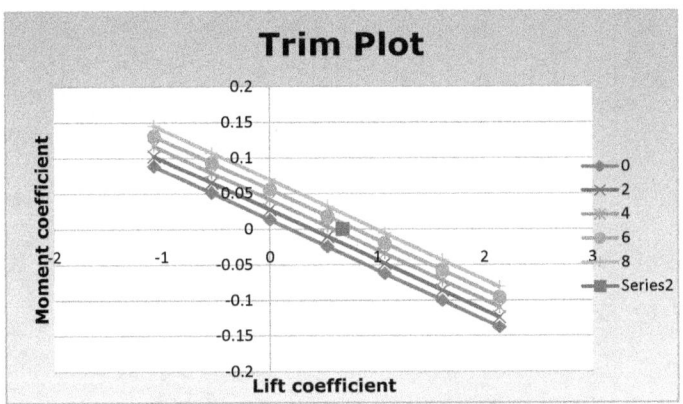

[Fig 11.5: Moment Coefficient *Vs* Lift Coefficient]

Chapter 12 | Propulsion

12.0 Propulsion

12.1 Introduction

Propulsion is the act of propelling a vehicle. Whether it is a ground vehicle or an aircraft, a propulsive unit is must required for the motion of the body and to perform the required mission for which the aircraft is designed. Based on the requirements of the flying vehicle, there are different categories of the propulsion systems namely *Reciprocating engine/propeller, Turbojet, Turbofan,* and *Turboprop.* In each case, the primary concern is with two characteristics of these systems- thrust (or power) and the fuel consumption. These are the two propulsion quantities that directly dictate the performance of the aircraft. The thrust produced by a propulsion system and the efficiency with which it operates depends upon the type of the propulsion system. It can be stated that a propeller/reciprocating engine combination produces comparably low thrust with great efficiency and a turbojet produces considerably higher thrust with less efficiency. Therefore, there is a tradeoff between the two variables and that is the reason why all types of propulsion systems are in use today.

Designing a solar powered aircraft requires an electric motor to propel it. As batteries are the power source of these motors, therefore operating current and voltage, and efficiency of the motor are the main concerns for designing of a propulsion system besides thrust. A strategy was developed for the selection of the propulsion system for SAIST which involved selection of a motor, selection of a battery, selection of a propeller, and then optimizing the whole propulsion system in order to get the maximum performance of the propulsion system. The basic selection of the type of the motor and the battery was done by creating a decision criteria having figures of merit and selecting the highest weight choice for further analysis in the MotoCalcv8.0 software. The choices were then optimized by the software to get the optimized propulsion system for SAIST.

12.2 Electric Motors

An electric motor uses electrical energy to produce mechanical energy. The principle of operation of an electric motor is the reaction of the two magnetic fields upon each other within certain prescribed area. One of the fields is produces by an electrical current flowing in the motor windings while the other field is produced by a permanent magnet assembly. These two fields result in a torque which tends to rotate the rotor. As the rotor turns, the current in the windings is commutated to produce a continuous torque output.

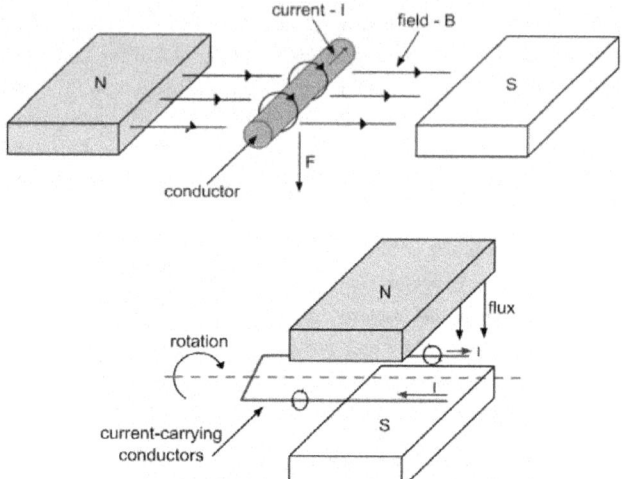

[Fig 12.1: Electric motor working principle]

12.2.1 Classification of Electric Motors

The electric motors are classified according to the type of the current that flows in the windings of the motor. According to this classification, there are two types of electric motors- *Alternating Current (AC) electric motors* and *Direct Current (DC) electric motors.*

- *Alternating Current (AC) electric motors:* These motors are driven by an alternating current. There are two basic parts of an AC motor- an outside stationery stator having coils supplied with alternating current to produce a rotating magnetic field, and an inside rotor attached to the output shaft which is given a torque by the rotating field.
- *Direct Current (DC) electric motors:* These motors are driven by a Direct current. The mechanism for producing the torque in the output shaft is the

same as the AC motor. The difference is the type of current that flows through the coils.

When selecting motor type for aerospace applications, careful attention is given to the specific application of electric motors. The factors that are considered essential for selection of an electric motor used for aerospace applications are application, environment, thermal, efficiency, weight, volume, life, complexity, torque, speed, power source, envelope, duty cycle, and controllability. DC motors have proven to be more successful for use in airplanes because they result in a light weight propulsion system and reduce the complexity; giving maximum efficiency at the same time. AC motors have a complex mechanism that constitute a heavy propulsion system overall. Therefore, in the coming analysis of the propulsion system, DC motors have been analyzed for the development of a propulsion system for a solar powered HALE aircraft.

12.2.2 Types of DC motors

By far the most common DC motor types are the *brushed* and *brushless* types, which use internal and external commutation respectively to create an oscillating AC current from the DC source - so they are not purely DC machines in a strict sense.

12.2.2.1 Brushed DC motor (BDC)

The Brushed Direct Current (BDC) motors use commutator and carbon brushes to apply current through the windings as the motor rotates. The BDC motor utilizes wound elements in the rotor and permanent magnets attached to a stationary stator ring. In a BDC motor, electrically separated motor windings are connected to the commutator ring. The commutator and brush-system form a set of electrical switches, each firing in sequence, such that electrical-power always flows through the armature-coil closest to the stationary stator (permanent magnet). Current is carried by spring loaded brushes, through the commutator into the windings of the rotor. The current in the windings creates magnetic fields, which react with the stator's permanent magnetic field. The magnetic repulsion causes the rotor to rotate. This rotation causes the brushes to make and break connections through the commutator with different winding pairs. The moving magnetic field provides the torque necessary to rotate the motor's armature. These types of motors are very inefficient typically in the region of 50-65%.

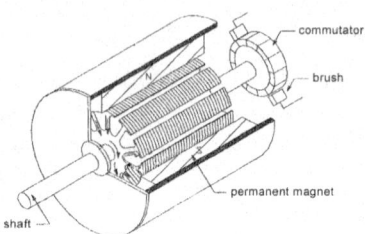

[Fig 12.2: Brushed DC motor]

The following table discusses the advantages and disadvantages of a brushed DC motor:

Advantages	Disadvantages
Low cost	Brush Dust
Simplicity	Brush to commutator arcing and wear
Availability	Electromagnetic interference
	Mechanical noise
	Short motor life
	Low efficiency
	Limited speed
	Poor thermal characteristics in vacuum

[Table 12.1: Brushed DC Motor advantages & disadvantages]

12.2.2.2 Brushless DC motors (BLDC)

The Brushless Direct Current (BLDC) motor uses electronic commutation to control the current through the windings. The BLDC motors use permanent magnets on the rotor. The BLDC motor contains rotor position sensor electronics so that the power input wave form to the windings is in sequence with the proper rotor position. Motor efficiency is enhanced because there is no power loss in the brushes. In the BLDC motor, the stator is wound with electromagnetic coils that are connected in a multiphase configuration, which provides the rotating field, and the armature consists of a soft iron core with permanent magnet poles. Sensing devices define the rotor position. The commutation logic and switching electronics convert the rotor position information to the correct excitation for the stator phases. Sensing devices include hall-effect transducers, absolute encoders, optical encoders, and resolvers. The electronic controller can be separate or packaged with the motor.

The direction of rotation can be changed by swapping any two of the three wires on the motor. Brushless motors are a lot more efficient generally in the range of 85-92% which means more of the power is converted to rotation and less to heat. This higher efficiency allows them to handle higher amp/power and have longer flight time. The following table discusses the advantages and disadvantages of a brushless DC motor:

Advantages	Disadvantages
High speed (up to 100,000 RPM)	Higher electronic cost
High torque at high speed	Greater motor drive complexity
Approximately double the output torque over BDC motor of the same size	
Windings on the stator instead of the rotor improves heat dissipation	
Long life due to the absence of the brushes	
High efficiency	
Vacuum compatible	

[Table 12.2: Brushless DC Motor advantages & disadvantages]

12.2.2.3 Comparison of Brushed and Brushless DC motors

The following table summarizes the comparison between both technologies:

Feature	Brushed	Brushless
Commutation	Mechanical	Electronically
Maintenance	High	Low
Electrical Noise	High	Low
Life	Shorter	Longer
Speed/Torque Characteristics	Moderately flat	Flat. (Enables operation at all speeds)
Efficiency	Medium	High
Motor Size	Larger due to commutator and difficulty removing heat	Smaller
Speed ranges	Commutator limits speed	Can rotate high speeds
Audible noise	High at high speeds because of brushes	Low
Drive Complexity	Simple and inexpensive	Complex and expensive
Control Requirements	No controller is required for fixed speed	A controller is always required to keep the motor running

[Table12.3: Brushed & Brushless Motors Comparison]

Besides the advantages shown in the table, brushless motor's commutation control can also easily be separated and integrated into other required electronics, thereby improving the effective power-to-weight and/or power-to-volume ratio.

It can be seen from the comparison table that Brushless DC motor has far better features than those of a Brushed DC motor.

12.2.3 Types of Brushless DC motor

12.2.3.1 Out-runners

Out-runner motors are sometimes referred to as 'rotating can' motors. In the design of the out-runner motor, the back part of the motor (where the wire comes out) is stationery and only the outer can and shaft rotates. Out-runners are suited to high torque and lower RPM applications. They throw large propellers at low RPM which are a lot more efficient than smaller propellers with high RPMs.

[Fig 12.3: Out-runner DC motor]

12.2.3.2 In-runners

In-runner motors are the type of brushless motors in which the can is stationery while the shaft spins. In-runner motors are suited to high speed low torque applications, geared applications and in electric helicopters.

[Fig 12.4: In-runner DC motor]

The features which have a greater weight for the motor selection for a HALE aircraft are *torque characteristics, efficiency, motor size,* and *RPM.*

These features are taken as figure of merits and given weights according to their importance and a decision matrix is formed for the final selection of the type or motor as follows:

Figure of Merit	Weight	Brushed	Brushless	Out-runner
Size (weight)	0.30	-1	1	0
RPM	0.25	0	0	1
Torque	0.30	0	1	0
Efficiency	0.15	-1	0	1
Total	**1**	**-0.45**	**0.60**	**0.40**

[Table 12.4: motor type down select]

Brushless DC motors are the most popular motor choice now in the electric aircraft industry. With their superior power to weight ratios, a large range of sizes, from the sub micro motors weighing in at less than 5 grams, to large motors producing power in the region of thousands of watts, they have revolutionized the radio control market. Their introduction has created a boom in the electric aircraft industry.

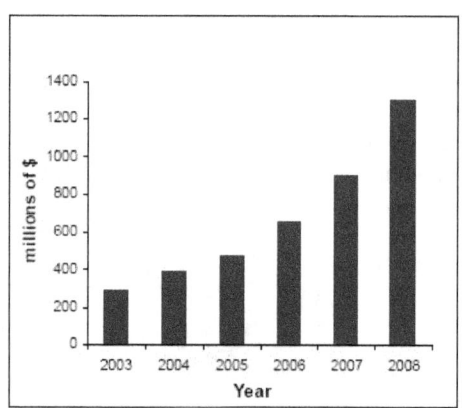

[Fig 12.5: Brushless DC motor Market Growth (source: ARC Advisory Group)]

12.3 Power requirement of the aircraft

The thrust requires by the HALE at different velocities for an un-accelerated flight is equal to its drag at those velocities and is given by the relations below:

$$TR = Drag = q_\infty S C_D = \frac{W}{C_L/C_D}$$

$$Parasite\ TR = q_\infty S C_{DO}$$

$$Induced\ TR = q_\infty S K C_L^2$$

This relation is plotted versus velocity to get the following curve:

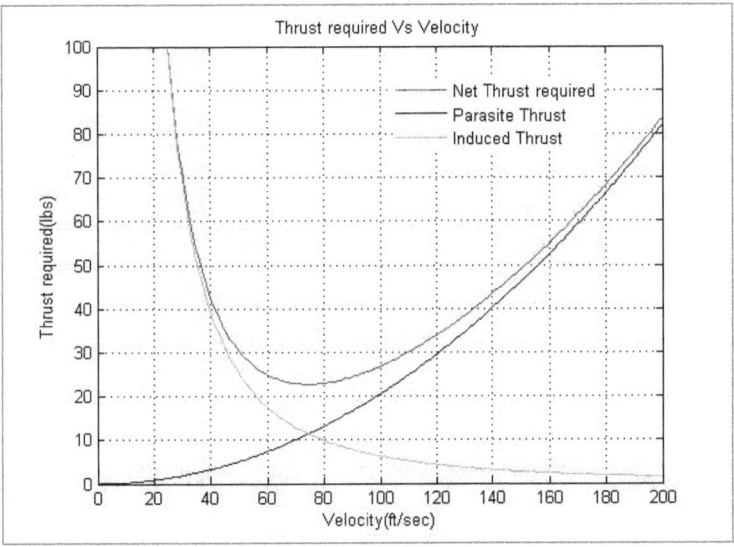

[Fig 12.6: Thrust required versus Velocity]

As the design point velocity is the loitering velocity of the aircraft which is 74 ft/s, the thrust required at the velocity is equal to 22.84 lbs. The thrust required is converted to Power required of the HALE by the relation given as:

$$P_R = T_R \times V_{inf}$$

P_R = 22.84 × 74 = 1690.32 lb-ft/s= 2291.766 Watts = 3.073 hp

12.4 Motors setup configuration

Prior to the final selection of the motor, it is required to find out the motors configuration on the aircraft whether to use a single motor power or a distributed power from a number of motors. For that purpose the following points were considered:

- As the aircraft has a large span of 86ft, the use of a single motor configuration would require a motor of very high power. The size of the motor would then be accordingly greater and more load of the propulsion system would be concentrated at one point on the aircraft which would require more structural strength at that point.
- A single motor configuration can also cause mass balancing problems.
- A very large propeller would be required to produce the required thrust out of the motor which would need the aircraft to be high above the ground and consequently require larger size of the landing gear supporting panels.
- The larger size of the landing gears would create more drag and thus more thrust will be required.
- The greater the distribution of the stream of the air coming out of the propellers, the efficient would be the control power of the control surface.

Keeping in view all of the above facts, a distributed configuration is used for the motor placement on the aircraft.

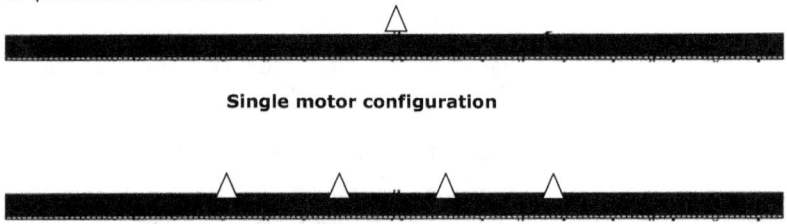

Single motor configuration

Distributed configuration

[Fig 12.7: Motors Configuration Comparison]

Keeping in view the large span of the aircraft and the reference aircraft, the number of motors to be employed on the aircraft is chosen to be four.

12.5 Selection of Motor

The power of a motor is selected on the basis of the required altitude to be achieved by the aircraft as well as an appropriate rate of climb at the same time fulfilling the basic power requirement of the aircraft. Furthermore, as the motors are to be powered by the solar panels, therefore there is also a constraint of the power delivered by the solar panels to the motor. Keeping in view these facts, a few motors were selected for the analysis. The specifications of these motors were obtained and the power losses were calculated for each of the candidates. A decision matrix was generated having efficiency, weight, Kv values, current, and the power losses as figures of merit.

12.5.1 Candidate Motors

Based on their power, a number of motors were selected for the analysis. These included Hacker brushless A60 and A80 series, and NEU motors 2215F and 1527F series. The following table shows the data of these motors:

Motor	Volt	Amp	Kv (RPM/V)	Power*(Watts)	I_o^*	R_m^*
Hacker A60-18M	35	67.2	190	2351	1.8	0.027
Hacker A60-18L	42	67.3	149	2825	1.6	0.020
NEU 1527/1.5Y	45	47	850	3300	1.5	0.012
NEU 2215/3Y	62	80	480	5000	0.9	0.018

[Table 12.5: Motors Characteristics]

Power*= Peak power

I_o^* = Idle Current

R_m^* = Coil Resistance

The data given in the above table is ideal i.e. without any power losses while in reality there are power losses in every motor which affects the output power and the Kv values. The major power losses in a motor are discussed below and then power losses for each motor are calculated for improvement in the motor selection. The efficiency for each motor is calculated using the actual power output including the losses. The Kv values are also calculated using the losses.

12.5.1.1 Power Losses

Every motor has power losses in it which means that a motor consumes more power than it delivers at its shaft. The motor's Output Power is equal to the Input Power minus the Power Loss. Most of Power Loss is equal to the sum of the Copper Loss plus Iron Loss. These losses are calculated using the following relations:

$$\textbf{Copper Loss} = \text{Coil's Resistance } \textbf{R}_m \times \text{Input Current } \textbf{I}_{in}$$

$$\textbf{Iron Loss} = \textbf{V}in \times \text{Idle Current } \textbf{I}o$$

The following equation can also be used to calculate the motor's output power:

$$\textbf{P}out = (\textbf{V}in - \textbf{I}in \times \textbf{R}m) \times (\textbf{I}in - \textbf{I}o)$$

12.5.1.2 Efficiency

Efficiency is a measure of how much of the Input Power (the power that the battery delivers to the motor) is actually used to turn the propeller (Output Power) and how much is wasted as heat.

The motor's Efficiency (η) is the ratio of the Output Power to the Input Power:

$$\% \ \eta = 100 \times \textbf{P}out / \textbf{P}in$$

12.5.1.3 Trend of efficiency with increasing voltage and current

Keeping the current constant first, increasing the voltage increases the efficiency until RPM limit is reached. The following graph shows the increasing efficiency of AVEOX 1406/3Y motor:

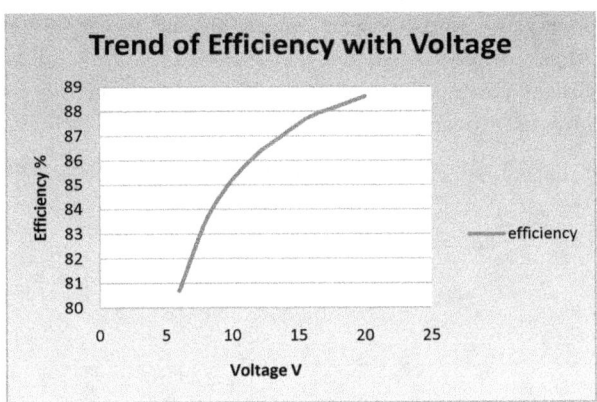

[Fig 12.8: Efficiency versus Voltage]

The above graph is drawn by keeping the current constant.

By keeping the voltage constant this time and increasing the current, the efficiency increases until the point where the RPM loss is equal to the torque loss. The following curve shows the trend:

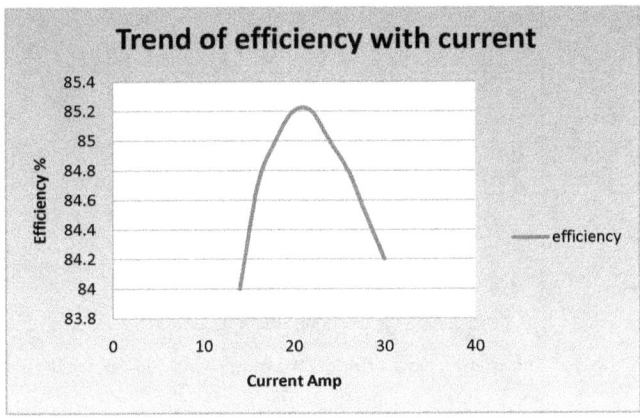

[Fig 12.9: Efficiency versus current plot]

12.5.1.4 Power losses and Efficiency Calculations for the Candidate motors

The power losses and efficiencies were calculated for each of the candidate motor using the relations described above. The final output power delivered by the motor is obtained by subtracting the power losses from the ideal power.

Motor	Weight(oz)	Loss (Watts)	Output Power(Watts)	Efficiency (%)
Hacker A60-18M	26.8	181.661	2170.338	92.2
Hacker A60-18L	32.1	155.632	2670.968	94.4
NEU 1527/1.5Y	28.6	93.162	2021.838	95.5
NEU 2215/3Y	33.0	169.704	4790.296	96.5

[Table 12.6: Motors Characteristics]

12.5.2 Final Selection of the Motor

The final selection of the motor was made by making a decision matrix having Efficiency, Weight, Kv values, Current and Power output Figures of Merit.

Figure of Merit	Weights	A60-18M	A60-18L	NEU 1527/1.5Y	NEU 2215/3Y
Efficiency	0.20	0	0	1	1
Weight	0.35	1	-1	0	-1
Kv	0.10	0	-1	1	0
Current	0.15	0	0	1	-1
Power output	0.20	1	0	0	-1
Total	1.0	0.55	-0.45	0.45	-0.45

[Table 12.7: FOM Comparison of Motors]

Based on the decision matrix, *Hacker A60-18M* motor was chosen for propelling the aircraft.

12.6 Battery Selection

The selection of a right battery is very important for the aircraft's performance. The primary consideration for the selection of the right battery is its capacity to store and retain charge at the same time having less weight. Three types of batteries were considered- Nickel Cadmium (NiCd), Lithium ion (Li-ion) and Lithium Polymer (Li-poly). Based on their advantages and disadvantages, the final selection was made.

Battery	Advantages	Disadvantages
Ni-Cd battery	<ul><li>Fast and simple charging.</li><li>Higher number of charging/discharging cycles.</li><li>Good low temperature performance.</li><li>Available in wide range of sizes and performance.</li></ul>	<ul><li>Relatively low energy density compared with newer systems.</li><li>Has relatively high self-discharge needs recharging after storage.</li><li>The NiCd contains toxic metals.</li></ul>
Li-ion battery	<ul><li>They have a higher energy density than NiCd.</li><li>They also operate at higher voltages than other rechargeable, typically about 3.7 volts for lithium-ion vs. 1.2 volts for NiMH or NiCd.</li><li>Lithium-ion batteries also have a lower self discharge rate than other types of rechargeable batteries.</li></ul>	<ul><li>Lithium-ion batteries are more expensive than similar capacity NiMH or NiCd batteries.</li><li>They are much more complex to manufacture.</li></ul>
Li-poly	<ul><li>LiPo battery packs have the highest power to weight ratio than any other battery available.</li><li>These batteries pack an enormous amount of energy and are much lighter than the</li></ul>	<ul><li>They are relatively fragile, cost more and can burst into fire if improperly charged.</li><li>Have very short shelf life of two to three years.</li></ul>

other two types. • LiPo batteries keep a constant power output throughout the flight. • Retain charge significantly longer than NiCad or NiMH batteries when not being used	• Very sensitive to high temperatures.

[Table 12.8: Batteries Advantages & Disadvantages Comparison]

Lithium polymer battery was chosen for its light weight and greater charge capacity as compared to the other two.

12.7 Optimization of the Propulsion package

After the selection of motor and the battery, the propulsion system was optimized for the final selection of the number of cells of the battery and the propeller to achieve an overall good performance. The optimization was done using 'Moto-Calc 8' software. The optimization was based on the overall in-flight performance of the propulsion system. As the software is constrained by the weight of the aircraft, the in-flight performance analysis was done using a single motor with weight of the aircraft reduced four times. A flow chart depicting the process of optimizing the propulsion system is shown in figure:

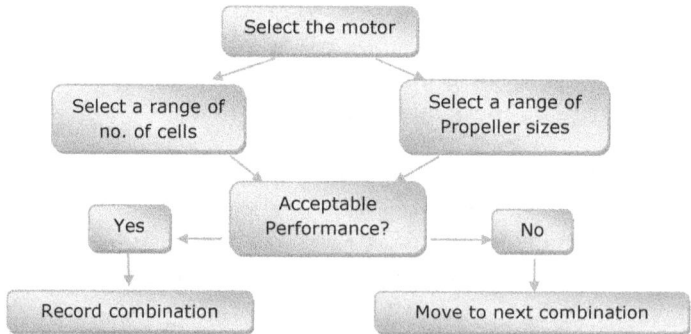

[Fig 12.10: Optimization of propulsion system flow chart]

First, the motor was selected in the software. The range of number of cells of the battery was selected and a range was selected for the propeller size too. The in-flight performance was analyzed using the software for each combination until the performance requirements were met. The optimized combination was then recorded and used for further analysis.

12.8 Propeller Selection

The propeller was selected from the combination that gave the maximum performance. The selected propeller size is **34″×10″**. The efficiency curve of the propeller with airspeed is shown below:

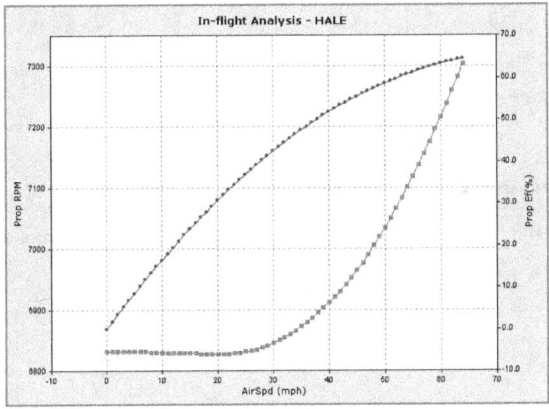

[Fig 12.11: Propeller efficiency versus airspeed]

The efficiency curve shows that the efficiency is close to maximum near the cruise velocity i.e. 61.3 miles/hour (90 ft/s).

12.9 Number of cells of the Battery

The number of cells selected from the optimized combination is 10 cells for a single battery. For four motors, four batteries would be used or a single battery with equivalent capacity of four will be used. The following graph shows the battery's in-flight performance.

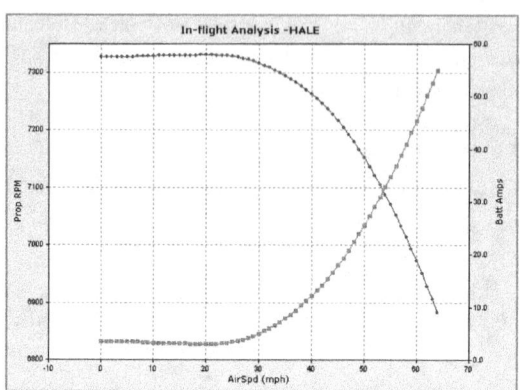

[Fig 12.12: Plot of Battery amps versus Airspeed]

12.10 Performance curves from the Optimized combination

Following are the performance curves obtained from the optimized configuration:

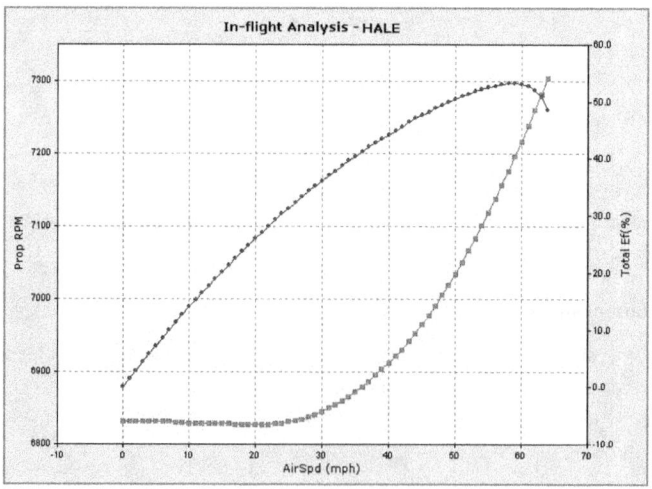

[Fig 12.13: Total Efficiency vs. Airspeed]

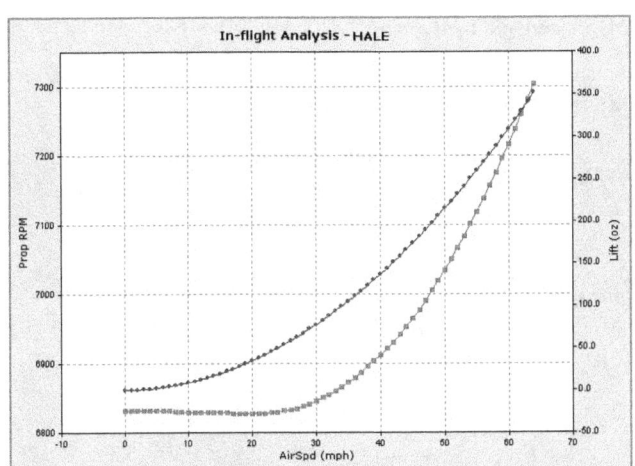

[Fig 12.14: Lift vs. Airspeed Plot]

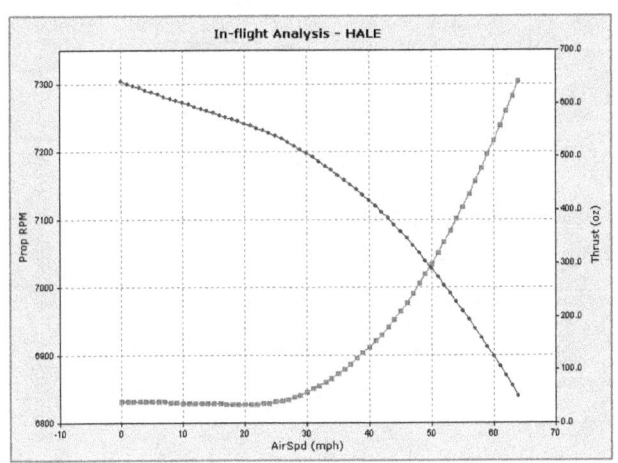

[Fig 12.15: Thrust vs. Airspeed Plot]

12.11 Final Propulsion System

After analyzing each component of the propulsion system and the performance of its combination, the following system was finalized.

Optimized Propulsion System	
Motor	Hacker Brushless A60-18M
Propeller	APC-E 34×10
Battery	LiPo 10 cell
Speed controller	Brushless ESC- 100 Amp

[Table 12.9: Final Propulsion System]

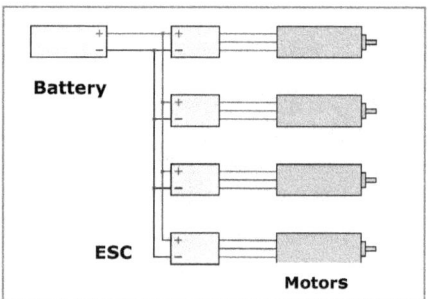

[Fig 12.16: Propulsion Circuit Without solar panels]

Chapter 13 | Solar Power System Design

13.0 Solar Power System Design

In order to make a high altitude and long endurance aircraft it is well understood that the aircraft requires a continuous power for long duration so with the available power system that mainly includes engines, batteries and fuel cells of other kind but the most economical and favorable power system for long endurance and high altitude flight is the solar power system that mainly involves the conversion of solar energy into electrical in order to drive the propulsion system. So designing a solar power source the following steps are considered and evolved as;

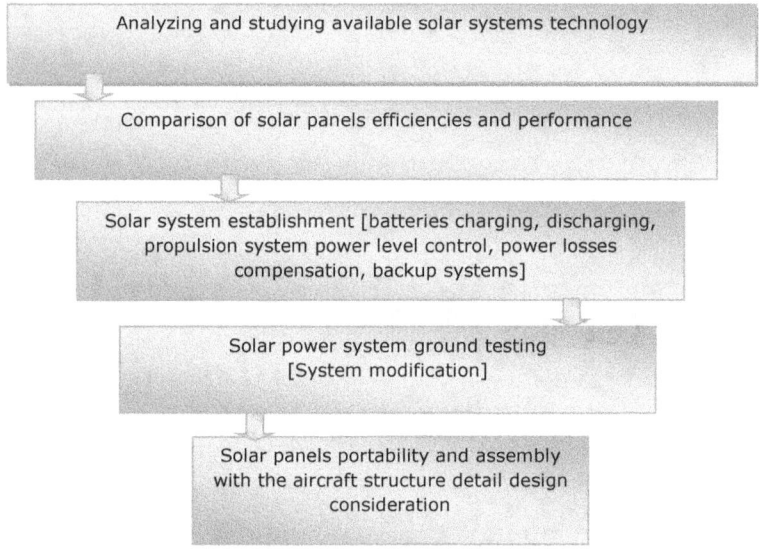

[Fig 13.1: Solar Power System Design Flow Chart]

Selection of baseline reference configuration depends upon the requirement and the evaluation of already existing solar powered high altitude long endurance configurations.

13.1 Power Management

Major challenge is the power management that has to ensure continuous flight by day and night. During the day, our aircraft will have to retrieve solar energy for continuous leveled flight, but also for scientific measurements, navigation and control electronics and battery charging for night flight. So, power management, from solar cells to the motor and the propeller is a main issue.

Solar power

At noon, the solar power at latitude lower than 20°N on Earth is **950 [W/m²]** during a sunny day as some assumptions are that of **1000 [W/m²]**. According to the variable position of the sun in the sky, the solar power function can be approximated by a cosine (Fig. 2) and the mean power on **24 hours** is estimated with:

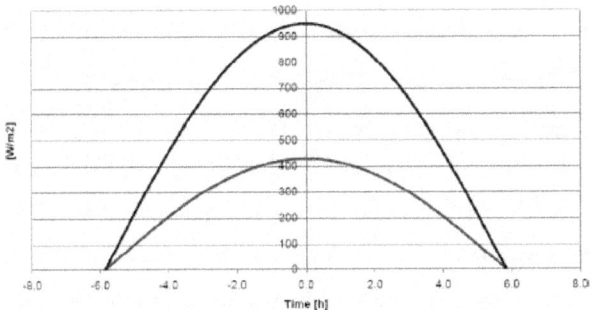

[Fig 13.2: Day Solar power on earth and mars]

So, taking average of irradiance using peak value of $950\frac{W}{m^2}$

$$P_{mean\_Earth} = \frac{P_{peak_{Earth}}}{\pi} = 302.54[\frac{W}{m^2}]$$

So, using values of power to area ratio values for earth as mean values we have the area of 528ft² or 49.05m² so, calculating power obtained if we use solar panels with maximum efficiency on the complete plan form then;

$$P_{obtained} = 302.54 \times 49.05 = 14839.96watts = 20hp$$

Where, power required in our case is 3.07hp and maximizing it, we have the power available requirement is that 6.52hp.

$$P_{required} = 3.07hp = 2289.299\ watt$$

$$P_{available} = 6.52hp = 4861.963 \text{ watt}$$

Battery

Battery selection is the most critical issue for an autonomous solar powered aircraft because it represents the most important part of the total weight. After having compared all energy storage technologies available today, Lithium-Ion- Polymer batteries have been selected for our project, because it represents currently the best energy/mass ratio that is commercially available.

The batteries are mainly used during the night flight, when the solar power is below the power requirement for a leveled flight. As the total effective night duration is 13.2 hours, we can calculate the needed capacity:

Battery Capacity = 13.2 [h] ×83.25 [W] =1098.9 [Wh]

Finally, including a margin of 20 [%], our battery is composed of five Li-polymer battery cells, which lead to a capacity of 1098.9 [Wh] with 4500 [mAh] at 18.5 [V] and a total weight of 0.62 [kg].

13.2 Solar Panels Design

Solar panels, composed by solar cells connected in a certain configuration, cover a certain surface of wing or other part of the airplane .During the day, depending on the sun irradiance and the inclination of the rays, the convert light into electrical energy. A converter, called Maximum Power Point Tracker, ensures that the maximum amount of power is obtained from the solar panels. This power is used firstly to power the propulsion group and the onboard electronics, and secondly to charge the battery with surplus of energy.

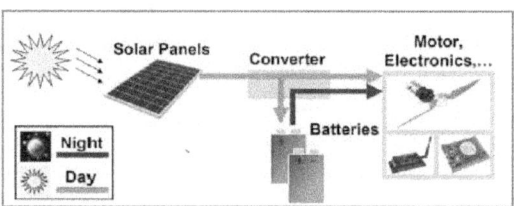

[Fig 13.3: Schematic Diagram of Solar Propulsion]

In the designing of solar mechanism the selection of solar panels is also a very critical part of the project as it involves the tradeoff between different parameters according to the design requirements and design constraints specially weight, power, voltages and surface areas. So, different commercially available solar panels have been compared in order to converge on the best and appropriate solar panels majorly fulfilling power requirements.

13.2.1 Solar Panels Model Characteristics

Solar Panels Models	Power (W)	Weight (lbs)	Area (ft²)	Voltage (V)	Current (A)
GSE-6	6	1.3	1.11	15.8	0.38
GSE-12	12	4.5	2.03	15.8	0.76
GSE-30	30	11	4.23	17.5	1.7
P3-30	30	1	0.619	12	1.5
P3-55	55	3.7	0.687	16	2.8
SL-6.5	6.5	0.45	0.312	12	0.43
SP-12	12	0.9	0.312	12	0.80
SL-25	25	1.7	6.15	12	1.5

[Table 13.1: Solar Panels Models Characteristics]

Selecting **P3-30 solar panel model** we have power calculations as:

Maximum numbers of solar panels placement available=528/0.619=853
Maximum Power obtained=853×30=25590watts
Maximum weight=853×1.4=1194.2lbs
Power Required=2289.299watts

So weight penalty will focus our design towards limited numbers of solar panels that is on $1/8^{th}$ of the total surface area.

Numbers of solar panels=106
Power outcome=106×30=3180watts
Weight=106×1=106lbs
Solar Panels Surface Area=106×0.619=65.614ft²

13.2.2 Energy Conversion Efficiency

A solar cell's energy conversion efficiency (η, "eta"), is the percentage of power converted (from absorbed light to electrical energy) and collected, when a solar cell is connected to an electrical circuit. This term is calculated using the ratio of the maximum power point, P_m, divided by the input light irradiance (E, in W/m^2) under standard test conditions (STC) and the surface area of the solar cell.

$$\eta = P_m/(EA_C)$$

P_m = Maximum Power Point (W)

E = Input Light Irradiance (W/m^2)

A = Surface Area of the solar cell (m^2)

$$\eta = 0.52$$

STC specifies a temperature of 25°C and an irradiance of 1000 W/m^2 with an air mass 1.5 (AM 1.5) spectrum. These correspond to the irradiance and spectrum of sunlight incident on a clear day upon a sun-facing 37°-tilted surface with the sun at an angle of 41.81° above the horizon. This condition approximately represents solar noon near the spring and autumn equinoxes in the continental United States with surface of the cell aimed directly at the sun.

13.2.3 Solar Cell Losses

The losses of a solar cell may be broken down into;

- Reflectance losses,
- Thermodynamic efficiency
- Recombination losses
- Resistive electrical loss

The overall efficiency is the product of each of these individual losses.

Due to the difficulty in measuring these parameters directly, other parameters are measured instead: Thermodynamic Efficiency, Quantum Efficiency, V_{OC} ratio, and Fill Factor. Reflectance losses are a portion of the Quantum Efficiency under "External Quantum Efficiency". Recombination losses make up a portion of the Quantum Efficiency, V_{OC} ratio, and Fill Factor. Resistive losses are predominantly categorized under Fill Factor, but also make up minor portions of the Quantum Efficiency, V_{OC} ratio.

13.2.4 Fill Factor

Another defining term in the overall behavior of a solar cell is the fill factor (FF). This is the ratio of the maximum power point divided by the open circuit voltage (V_{oc}) and the short circuit current (I_{sc}):

$$\eta = P_m/(V_{OC}I_{SC}) = \eta A_C E/V_{OC}I_{SC}$$

13.2.5 Maximum Power Point Tracker

A maximum power point tracker (or MPPT) is a high efficiency DC to DC converter which functions as an optimal electrical load for a photovoltaic (PV) cell, most commonly for a solar panel or array, and converts the power to a voltage or current level which is more suitable to whatever load the system is designed to drive.

The benefits of MPPT regulators are greatest during cold weather, on cloudy or hazy days or when the battery is deeply discharged. Solar MPPTs can also be used to drive motors directly from solar panels. The benefits seen are huge, especially if the motor load is continuously changing. This is due to the fact that the AC impedance across the motor is related to the motor's speed. The MPPT will switch the power to match the varying resistance.

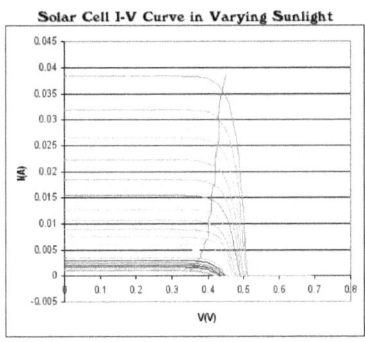

[Fig 13.4: Solar cell I-V curve]

Maximum Power Point Tracker (MPPT) is required to adapt the voltage of the solar panels so that they provide the highest power possible. Its mass is proportional to the maximum power it has to convert, which can be calculated using the solar panels area calculated above as showed in equation below. The constant k_{mppt} was found based on a study of existing high efficiency products.

$$m_{mppt} = k_{mppt} P_{solmax} = k_{mppt} I_{max} \eta_{cells} \eta_{mppt} A_{solar}$$

13.2.6 Power density of high efficiency MPPT's

The plot shown below approximate the maximum power output with reference to weight and has been drawn for different aircrafts, the trend seems to be linear.

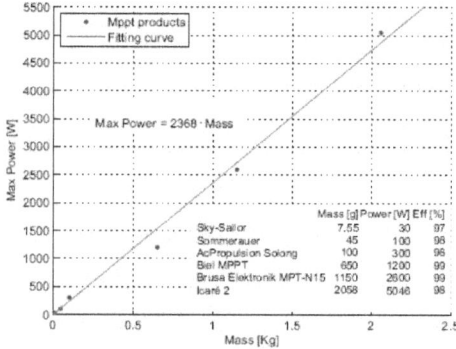

[Fig 13.5: Output Power versus weight plot for different solar aircrafts]

An MPPT controller, in addition to performing the function of a basic controller, also includes a DC voltage converter, converting the voltage of the panels to that required by the batteries, with practically no loss of power. In other words, they attempt to keep the panel voltage at their Maximum Power Point, while supplying the varying voltage requirements of the battery.

Manufacturers claim up to 40% power increase from your panels using MPPT, which is most likely to be achieved when battery levels are low and/or light levels are low.

13.3 Solar Power system Design

Solar power system layout for SAIST is as follows:

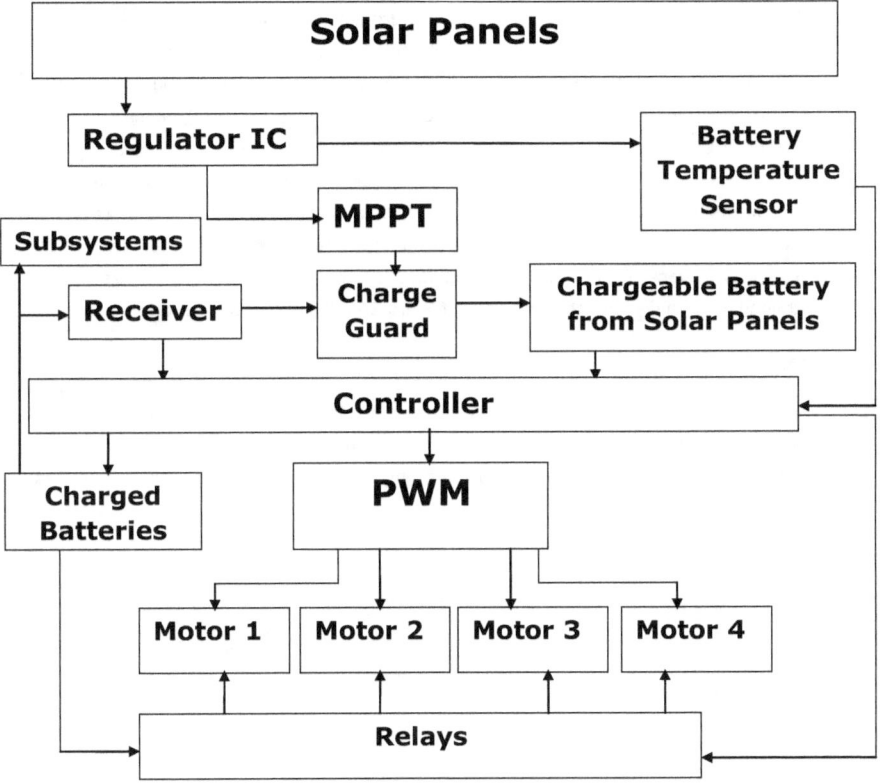

[Fig 13.6: Solar Power system in SAIST]

Thus above block diagram shows general solar power system for the complete aircraft in which solar energy is processed and feed to the propulsion system as due to limited efficiencies of solar panels and during night time a separate back up batteries system has also been established in order to complete the required power for 12 hrs flight.

Different components have been shown in the block diagram above in which regulator integrated chip has been used in order to regulate the voltage of the solar panel. MPPT is the maximum power point tracker the usage and application of which has been explained above. Battery temperature sensor regulates the temperature of the battery and charge guard is used with the battery to avoid overcharging in the battery. Controller is sending and receiving commands and operating accordingly. Where, along with motors pulse width modulation technique (PWM) is used in order to control the power loads and to monitor the speed of the motors. Relays are the ON-OFF switches that operate during varying power levels in the circuitry. Receiver and subsystems placed are powered by backup a battery that also powers the microcontrollers as well. Further, power from the solar panels have endurance of 10 hrs but with back up batteries and adding endurance increase of 3 to 5 hrs will be attained. So, finally with the tolerance of 3 hrs SAIST has 12hrs endurance.

13.4 Conclusion

After overview of solar power system requirement it is concluded that due to limited availability of solar panels efficiencies and other constraints it is more appropriate to implement solar power system and then predict the actual power output, efficiencies and performance.

Chapter 14 | Structural Design

14.0 Structural Design

This section involves the structural design of the aircraft that mainly includes the development of all flying wing configuration. Structural analysis has been done in order to predict the behavior of loading on the aircraft. Furthermore, different loading conditions have been predicted for making efficient structural design. This is not a detailed structural design but efforts have been made to make an approximate aircraft substructure configuration. The main components of the flying wing and the reinforcing components are considered and analyzed individually as well as the final assembly has been simulated for obtaining the most actual loading effects on aircraft during flight.

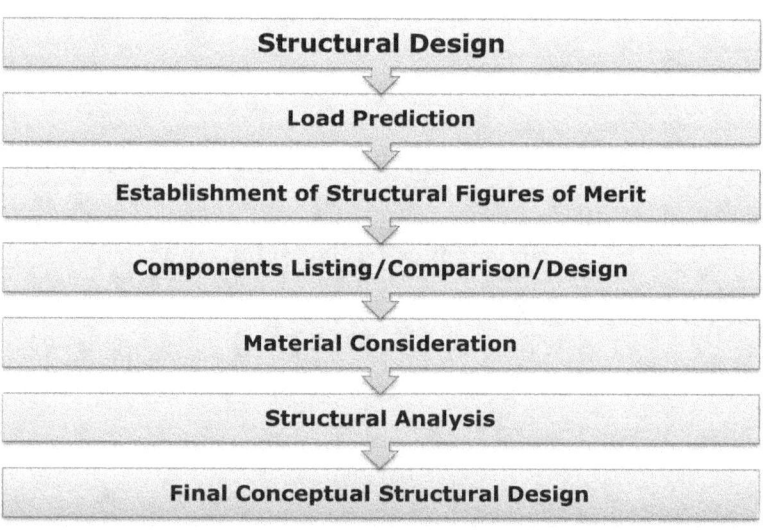

[Fig 14.1: Structural Design Sequence]

The main aircraft structural components are:

- Wing
- Spar
- Ribs
- Stringers
- Skin

14.1 Figures of Merit

Establishing the structural design criteria we have mainly considered the following variables in order to select and design the efficient sub structural components as all considerations of structural components are compared and evaluated across these figures of merit as follows:

- Weight
- Cost
- Reliability
- Manufacturing Processing

14.2 Load Predictions

There are mainly two types of loads that are aerodynamic loads and structural loads. Aerodynamic loads mainly involve lift loads and drag loads where structural loads are due to the loading conditions and weight.

14.3 Geometry Modeling

Geometry modeling of different structural components is as follows:

14.3.1 Wing

SAIST AHE UAV has a flying wing configuration so structural design configuration of the wing is more critical as it's the only surface of the aircraft. The substructure of the wing has two rectangular spars with the forward spar at 20 % chord and the aft spar at 75 % chord.

14.3.2 Spar

The substructure selection evaluates the number of spars, spar cross section, and spar material. The number of spars are evaluated and compared on the basis of figures of merit defined for structural design.

- One main spar Concept
- Two Spars Concept
- Three Spars Concept

A figure of merit comparison is shown below;

FOM	Weights	One Spar	Two Spar	Three Spar
Weight	4	3	2.5	1.5
Cost	3	2.5	2	1
Reliability	4	1	3.5	3.5
Manufacturing	2	2	2	1
Total	13	8.5	10	7

[Table 14.1: FOM Comparison]

The one spar concept costs slightly less to produce because of a reduction in manufacturing complexity but loading conditions and structure cannot withstand with one spar and results in twisting or warping in the structure during flight. So, additional structural reinforcement will be required to increase torsional rigidity.

The two spar concept has superior aerodynamic qualities compared to the other concepts as it is the most commonly used structural concept capable of withstanding loads according to the load distribution between main and secondary spars and further there are no interference drag penalties that result large number of additional fasteners required to connect a third spar.

The three spar concept involves weight and cost penalty and further structural performance in this case is much more reliable but it's not one of the optimized structural design options.
So, finally a two spars concept has been selected for HALE wing configuration finally
There are four types of spar cross sections most commonly used which are;

- Circular
- I-type
- Box-type
- C-type

All these cross sectional areas are compared and evaluated according to the figures of merit. We have selected I-type cross section for our HALE wing spar.
The main spar is located at approximately 20 % chord. The aft spar is located at approximately 75 % chord.

14.3.3 Ribs

Ribs the main load carrying component after spars in a wing the strength of the rib is very crucial for the wing under loading. The thickness of the ribs is 0.0164ft. The ribs caps have an area of 0.0164 ft^2.The total number of ribs placed along the wing are 83.

14.3.4 Stringers

There are three stringers spaced evenly among the top skin, and two stringers spaced evenly along the lower skins. The stringers have an area of 0.000489ft^2. These stringers will be designed such that they will appear continuous along the length of the wing.

14.3.5 Skin

Skin is also an important structural component and its weight is important for overall structural design. The thickness of the skins is 0.00175ft for the lower skins and 0.0025ft for the upper skins.

14.4 Modeled Geometry

After, designing structural components of SAIST the main components are modeled according to the dimensions calculated for spars and ribs.

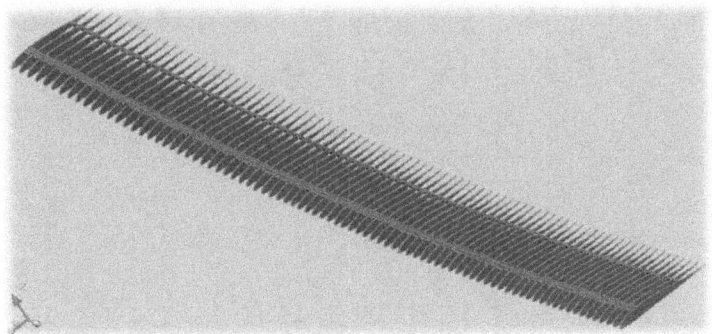

[Fig 14.2: SAIST Geometry Modeling]

[Fig 14.3: Spar and Ribs Assembly]

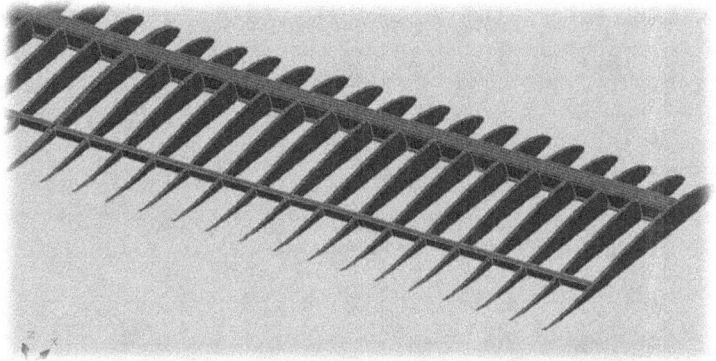

[Fig 14.4: Trailing Edge spar and ribs Assembly]

14.4.1 Complete Aircraft Assembly

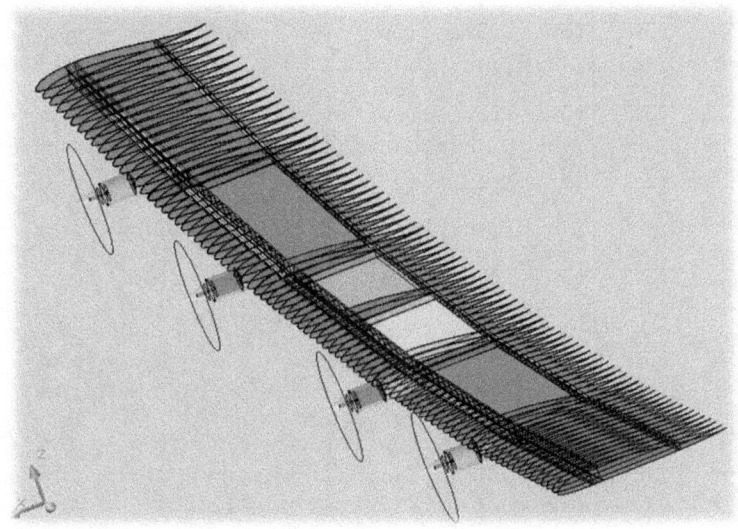

[Fig 14.5: Final Assembled 3D CAD Drawing]

Chapter 15 | Material Consideration

15.0 Material Consideration

Material selection for SAIST HALE UAV has been made on the basis of design consideration and the available technology of aerospace materials. Material selection is the most important consideration in the structural design in which after loading predictions and under different flight conditions, material for different parts of the aircraft is selected. Furthermore, reinforcement concept of using strengths of variable material at time is also one of the design considerations.

15.1 Manufacturing Variants

The selection of material is one of the most important aspects in structural design. For our aircraft we have the following material considerations as:

- Composites
- Aluminum/Alloys
- Wood

15.1.1 Composites

Composites are different from metals. They are combinations of materials differing in composition or form. The constituents retain their identities in the composites and do not dissolve or otherwise merge completely into each other although they act together. Reinforced concrete is an excellent example of a composite structure in which the concrete and steel still retain their identities.

In aircraft construction the term composite structures refers to fabric resin combinations in which the fabric is embedded in the resin but still retains its identity. Advanced composite materials consist of new high strength fibers embedded in an epoxy matrix. These composites provide for major weight savings in airplane structures since they have high strength to weight ratios. When replacing aluminum structure with graphite/epoxy composite weight reductions of 20% or better are possible. Weight reduction is the greatest advantage of composite material and is one of the key items in decisions regarding its selection. Other advantages over conventional structure include its high corrosion resistance and its resistance to damage from cyclic loading (fatigue). The major disadvantage of using advanced composite materials in airplane construction is the relatively high cost of the materials.

Composite Hybrids

Hybrids are made by the addition of some complementary material such as fiberglass or Kevlar to the basic carbon fiber/epoxy matrix. The added materials are used to obtain specific material characteristics such as greater fracture toughness and impact resistance, and should be considered for areas subject to foreign object damage.

The addition of carbon /epoxy to fiberglass structure is used to provide additional stiffness.

15.1.2 Aluminum Alloys

Aluminum is one of the most widely used metals in modern aircraft construction. It is vital to the aviation industry because of its high strength/weight ratio, its corrosion-resisting qualities, and its comparative ease of fabrication. The outstanding characteristic of aluminum is its light weight. In color, aluminum resembles silver, although it possesses a characteristic bluish tinge of its own. Commercially pure aluminum melts at the comparatively low temperature of 1,216°F. It is nonmagnetic, and is an excellent conductor of electricity.

Commercially pure aluminum has a tensile strength of about 13,000 psi, but by rolling or other cold-working processes, its strength may be approximately doubled. By alloying with other metals, together with the use of heat-treating processes, the tensile strength may be raised to as high as 96,000 psi, or to well within the strength range of structural steel.

Aluminum alloy material, although strong, is easily worked, for it is very malleable and ductile. It may be rolled into sheets as thin as 0.0017 inch or drawn into wire 0.004 inch in diameter. Most aluminum alloy sheet stock used in aircraft construction ranges from 0.016 to 0.096 inch in thickness; however, some of the larger aircraft use sheet stock that may be as thick as 0.0356 inch.

One disadvantage of aluminum alloy is the difficulty of making reliable soldered joints. Oxidation of the surface of the heated metal prevents soft solder from adhering to the material; therefore, to produce good joints of aluminum alloy, a riveting process is used. Some aluminum alloys are also successfully welded.

15.1.2.1 Aluminum alloys in Aircraft Structures

Heat treatable Alloys

Heat-treatable alloys commonly used in aircraft construction (in order of increasing strength) which are mostly 6061, 6062, 6063, 2017, 2024, 2014, 7075, and 7178.

- Alloys 6061, 6062, and 6063 are sometimes used for oxygen and hydraulic lines and in some applications as extrusions and sheet metal.
- Alloy 2017 is used for rivets, stressed-skin covering, and other structural members.
- Alloy 2024 is used for airfoil covering and fittings. It may be used wherever 2017 is specified, since it is stronger.
- Alloy 2014 is used for extruded shapes and forgings. This alloy is similar to 2017 and 2024 in that it contains a high percentage of copper. It is used where more strength is required than that obtainable from 2017 or 2024.
- Alloy 7178 is used where highest strength is required. Alloy 7178 and 7075 contain a small amount of chromium as a stabilizing agent.

Non Heat treatable Alloys

Non heat-treatable alloys used in aircraft construction are 1100, 3003, and 5052.

- Alloy 1100 is used where strength is not an important factor, but where weight, economy, and corrosion resistance are desirable. This alloy is used for fuel tanks, fairings, oil tanks, and for the repair of wing tips and tanks.
- Alloy 3003 is similar to 1100 and is generally used for the same purposes. It contains a small percentage of manganese and is stronger and harder than 1100, but retains enough work ability that it is usually preferred over 1100 in most applications.
- Alloy 5052 is used for fuel lines, hydraulic lines, fuel tanks, and wing tips. Substantially higher strength without too much sacrifice of workability can be obtained in 5052. It is preferred over 1100 and 3003 in many applications.
- Alclad is the name given to standard aluminum alloys that have been coated on both sides with a thin layer of pure aluminum. Alclad has very good corrosion-resisting qualities and is used exclusively for exterior surfaces of aircraft.

15.1.3 Wood

Wood is the third manufacturing variant in SAIST UAV but due to huge span and availability of light weight and high strength materials as in the case of composites and aluminum alloys the wood consideration is ignored on the vital structural components design of the SAIST.

15.2 SAIST Material selected

So, finally for SAIST different material type considered for main components of the structure is as;

SAIST Main Structural Components	Material Selected
Spars	2024-T3 Aluminum Extrusion
Ribs	Aluminum alloy 2024/2014
Skin	Carbon/Epoxy
Stringers	Carbon/Epoxy Composites
Reinforcement Joints	Hybrid Composites
Mounts	Aluminum alloy 7178
Wing Tips	Aluminum alloy 5052/1100

[Table 15.1: SAIST Material Selection]

The characteristics of the material considered have been fed into the FEM analysis software for structural analysis.

Chapter 16 | Structural Analysis

16.0 Structural Analysis

Structural analysis is probably the most common application of the finite element method. The term *structural* (or *structure*) implies not only civil engineering structures such as bridges and buildings, but also naval, aeronautical, and mechanical structures such as ship hulls, aircraft bodies, and machine housings, as well as mechanical components such as pistons, machine parts, and tools.

16.1 Types of Structural Analysis

There are many types of structural analyses few of them are described briefly below;

- *Static Analysis*--Used to determine displacements, stresses, etc. under static loading conditions. Both linear and nonlinear static analyses. Nonlinearities can include plasticity, stress stiffening, large deflection, large strain, hyper elasticity, contact surfaces, and creep.
- *Modal Analysis*--Used to calculate the natural frequencies and mode shapes of a structure. Different mode extraction methods are available.
- *Harmonic Analysis*--Used to determine the response of a structure to harmonically time-varying loads.
- *Transient Dynamic Analysis*--Used to determine the response of a structure to arbitrarily time-varying loads. All nonlinearities mentioned under Static Analysis above are allowed.
- *Spectrum Analysis*--An extension of the modal analysis, used to calculate stresses and strains due to a response spectrum or a PSD input (random vibrations).
- *Buckling Analysis*--Used to calculate the buckling loads and determine the buckling mode shape. Both linear (eigenvalue) buckling and nonlinear buckling analyses are possible.
- *Explicit Dynamic Analysis*--Explicit dynamic analysis is used to calculate fast solutions for large deformation dynamics and complex contact problems.

16.2 Static Structural Analysis

A static analysis calculates the effects of *steady* loading conditions on a structure, while ignoring inertia and damping effects, such as those caused by time-varying loads. A static analysis can, however, include steady inertia loads (such as gravity and rotational velocity), and time-varying loads that can be approximated as static equivalent loads (such as the static equivalent wind and seismic loads commonly defined in many building codes).

Static analysis determines the displacements, stresses, strains, and forces in structures or components caused by loads that do not induce significant inertia and damping effects. Steady loading and response conditions are assumed; that is, the loads and the structure's response are assumed to vary slowly with respect to time. The types of loading that can be applied in a static analysis include:

- Externally applied forces and pressures
- Steady-state inertial forces (such as gravity or rotational velocity)
- Imposed (nonzero) displacements
- Temperatures (for thermal strain)
- Fluences (for nuclear swelling)

Finite Element Method

Finite element method models a structure as an assembly of elements or components with various forms of connection between them. Thus, a continuous system such as a plate or shell is modeled as a discrete system with a finite number of elements interconnected at finite number of nodes. The behavior of individual elements is characterized by the element's stiffness or flexibility relation, which altogether leads to the system's stiffness or flexibility relation. To establish the element's stiffness or flexibility relation, we can use the *mechanics of materials* approach for simple one-dimensional bar elements, and the *elasticity approach* for more complex two- and three-dimensional elements. The analytical and computational development is best effected throughout by means of matrix algebra.

16.3 SAIST Structural Analysis

SAIST HALE UAV structural analysis has been done in order to analyze the loading conditions of the aircraft as the most important component of the aircraft is the main load carrying component; spar so, its position and sizing is very critical in structural design. So, two different position of the main leading edge spar has been analyzed by simulating their structural analysis.

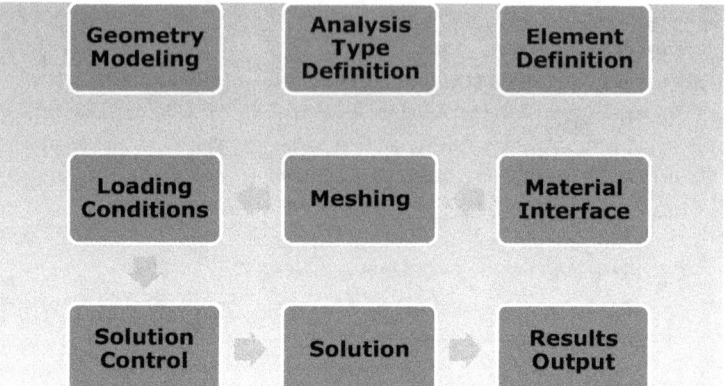

[Fig 16.1: Static Structural Analysis Processes in Ansys]

Geometry Modeling:

SAIST geometry has been modeled in CAD software package and has been transferred into Ansys.

Analysis Type Definition:

Analysis type definition involves selection of analysis type as in case of this project a structural analysis type has been defined using h-method.

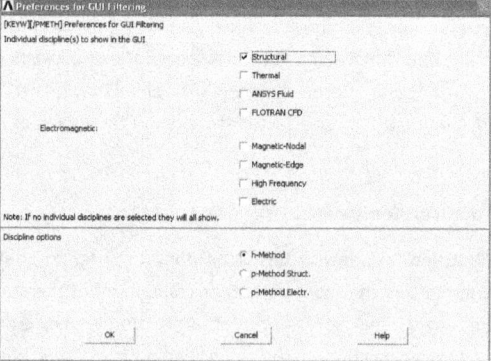

[Fig 16.2: Analysis type definition]

Element Definition

Element definition and selection according to particular structural geometry is a very critical part in structural analysis. So, in Ansys different types of elements are available, and for SAIST structural analysis SOLID186 and SOLID92 has been selected and their brief overview is as follows;

SOLID186: is a higher order 3-D 20-node solid element that exhibits quadratic displacement behavior. The element is defined by 20 nodes having three degrees of freedom per node (translations in the nodal x, y, and z directions). The element supports plasticity, hyper elasticity, creep, stress stiffening, large deflection, and large strain capabilities. It also has mixed formulation capability for simulating deformations of nearly incompressible elastoplastic materials, and fully incompressible hyperelastic materials.

SOLID92: has a quadratic displacement behavior and is well suited to model irregular meshes. The element is defined by ten nodes having three degrees of freedom at each node (translations in the nodal x, y, and z directions). The element also has plasticity, creep, swelling, stress stiffening, large deflection, and large strain capabilities.

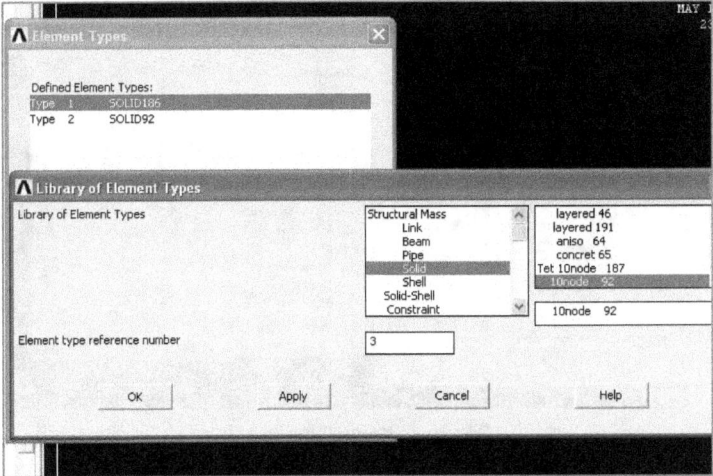

[Fig 16.3: Element Definition]

Material Model Interface

Material model interface involves the definition of material characteristics. In spar-rib configuration; structural analysis of SAIST involves Al-2024-T3 material having following characteristics;

Al-2024T3 Mechanical Properties	
Modulus of Elasticity	73.1GPa
Ultimate Tensile Strength	483MPa
Tensile Yield Strength	345MPa
Poisson's Ratio	0.33
Fatigue Strength	138MPa
Shear Modulus	28GPa
Shear Strength	283MPa

[Table 16.1: Material Characteristics]

So considering linear, elastic and isotropic material model behavior in SAIST structural analysis as;

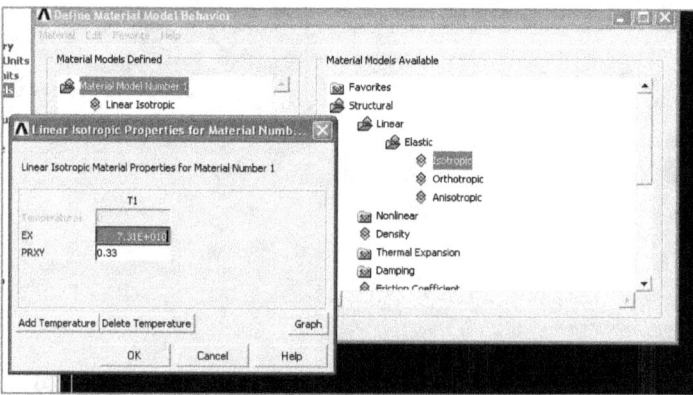

[Fig 16.4: Material Modeling]

16.3.1 Structural Analysis Approach:

SAIST HALE UAV structural design involves the determination of structurally favorable position and sizing of main leading edge spar that carries 80% of the main load. So, Structural analysis is done for design optimization of main spar configuration as;

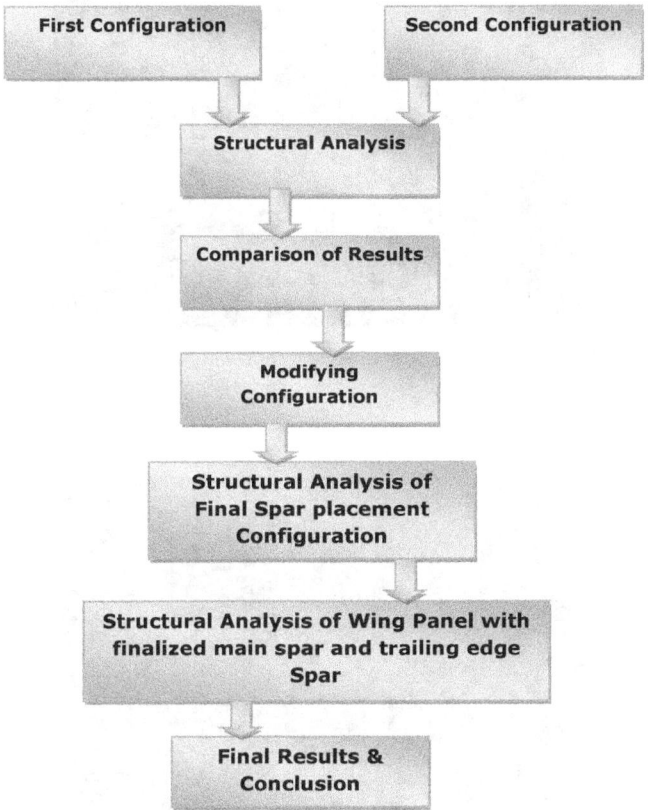

[Fig 16.5: Structural Analysis approach]

16.3.2 First Configuration:

First spar rib structural configuration involves spar placement at about c/4 location and the meshed geometry is as shown below;

Where loads are applied on first spar rib configuration as:

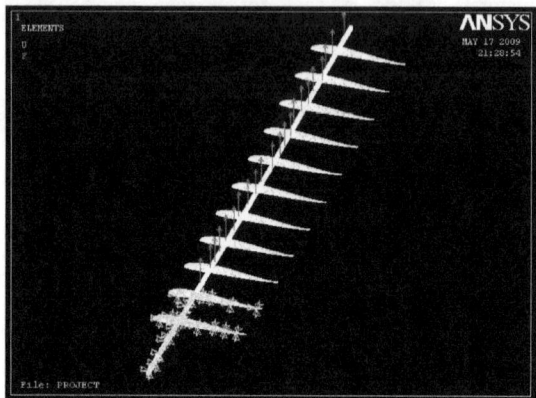

[Fig 16.6: Meshed Structure with spar]

16.3.2.1 Structural Analysis Results:

Deflection in X-Direction:

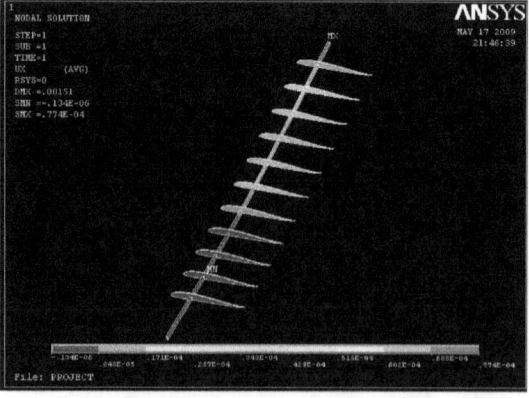

[Fig 16.7: Deflection in X-Direction]

Deflection in Y-Direction:

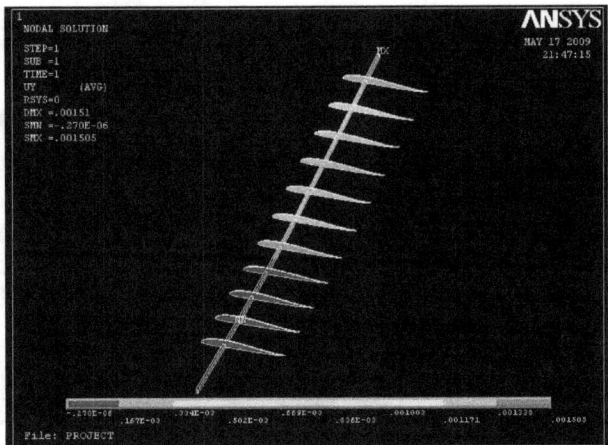

[Fig 16.8: Deflection in Y-Direction]

Deflection in Z-Direction:

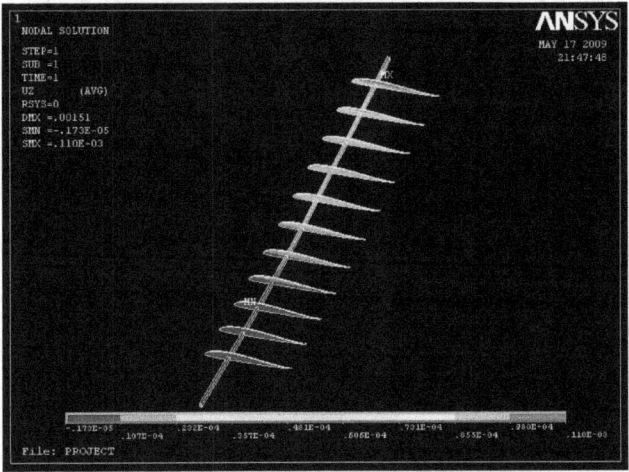

[Fig 16.9: Deflection in Z-Direction]

Total Deflection:

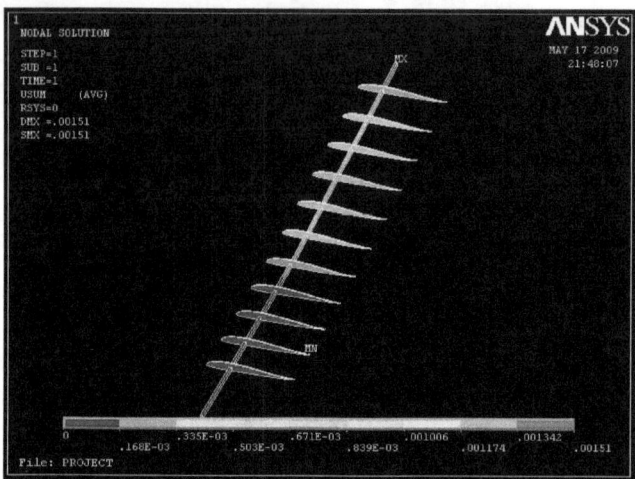

[Fig 16.10: Total Deflection in main spar]

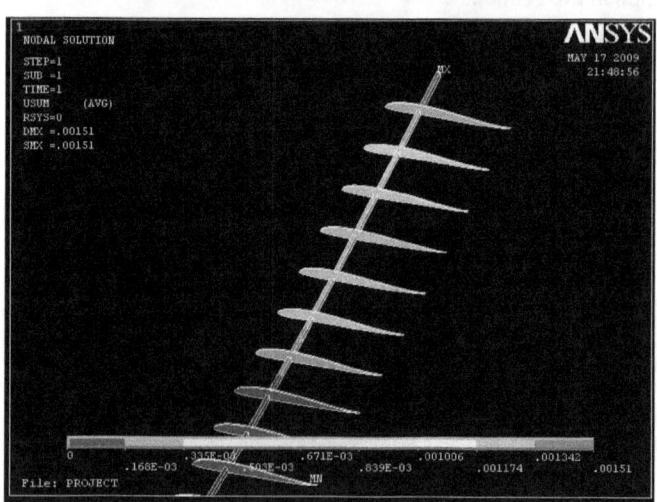

[Fig 16.11: Total Deflection in Structure]

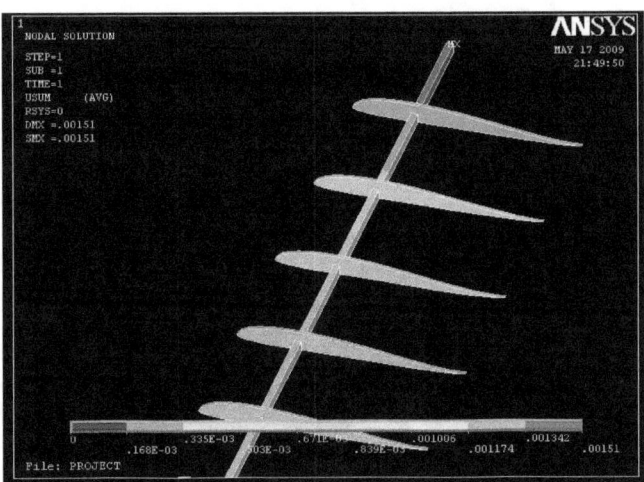

[Fig 16.12: Total Deflection in Main spar structure]

16.3.3 Second Configuration Analysis:

Second configuration has main leading edge spar placed after c/4 the analysis results are shown below;

Deflection in X-Direction:

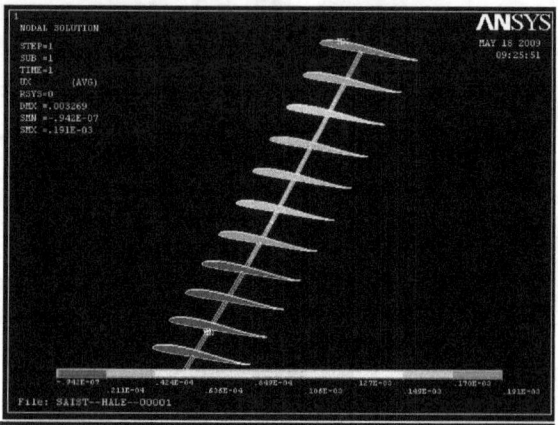

[Fig 16.13: Deflection in X-Direction]

Deflection in Y-Direction:

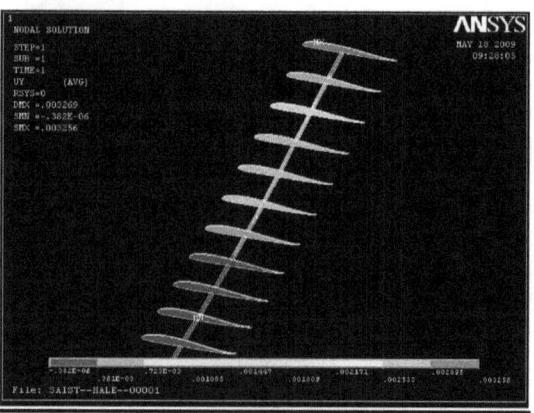

[Fig 16.14: Deflection in Y-Direction]

Deflection in Z-Direction:

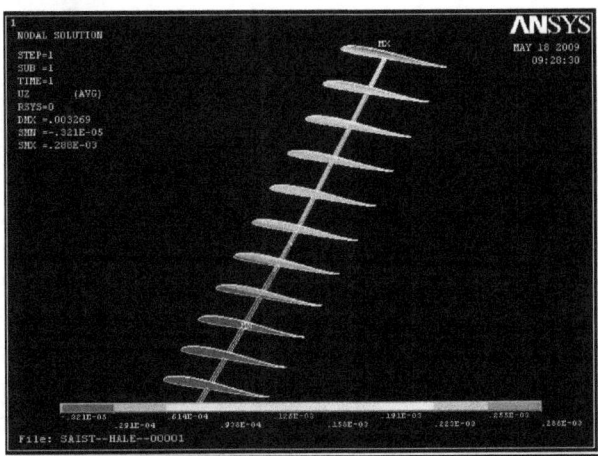

[Fig 16.15: Deflection in Z-Direction]

Total Deflection in Second Configuration:

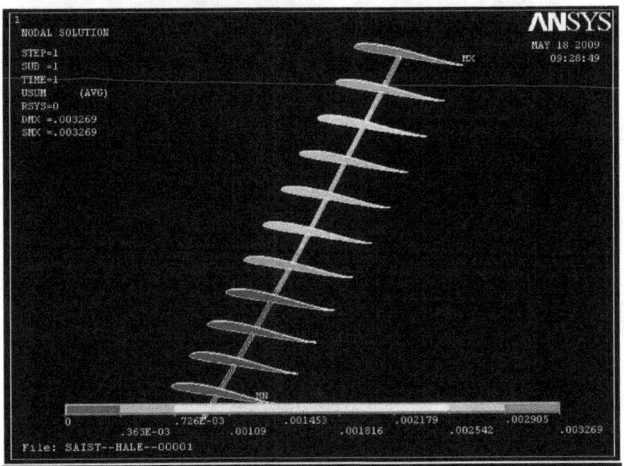

[Fig 16.16: Total Deflection in Structure]

16.3.4 Comparison of Results

Maximum total deflection comparison for both configuration shows first configuration is more favorable so location of main leading edge spar is finalized as just close to c/4 location from the leading edge.

16.3.5 Final Spar Configuration Analysis:

The dimensions and thickness is modified and finally spar configuration structural analysis has been made as follows;

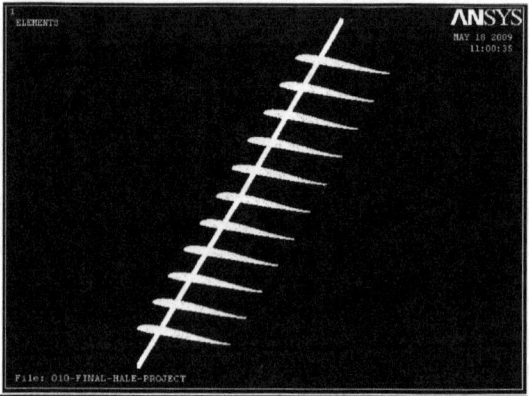

[Fig 16.17: Final Spar meshed Configuration]

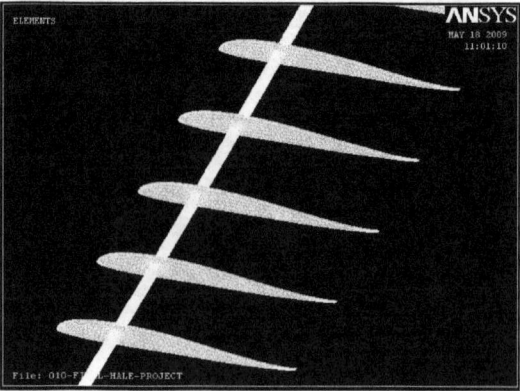

[Fig 16.18: Final Meshed Geometry]

16.3.6 Analysis Results:

Analysis results of deflection and deformation in structure are;

Deformation with undeformed Edges:

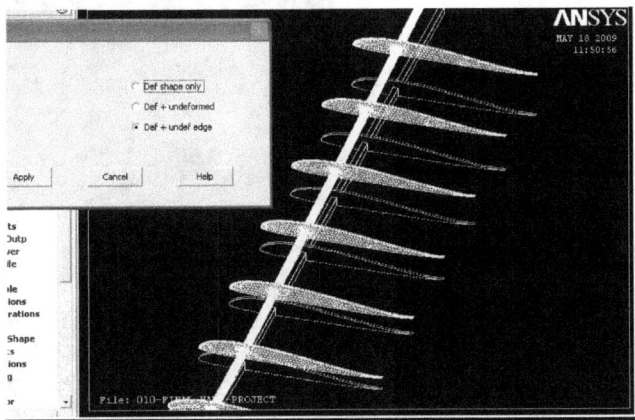

[Fig 16.19: Deformation with un-deformed Edges]

Deformation with undeformed Shape:

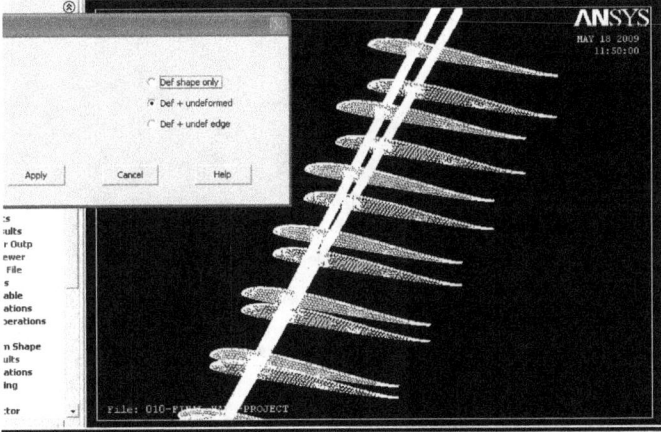

[Fig 16.20: Deformed with un-deformed Shape]

Deformed shape Output:

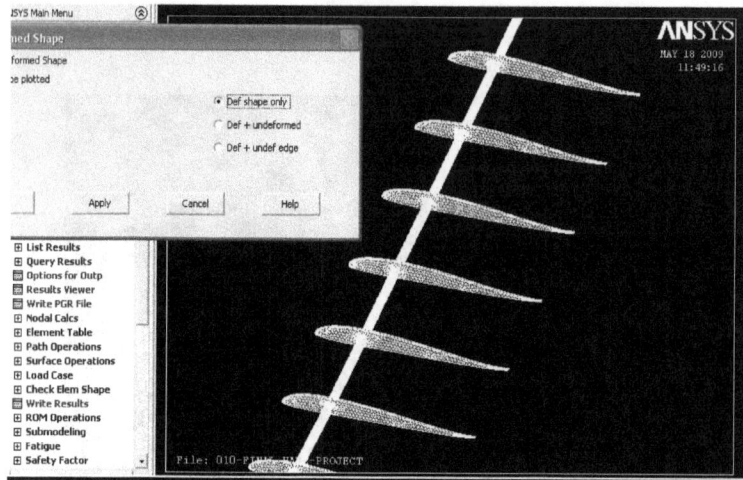

[Fig 16.21: Deformed Shape]

Deflection in X-Direction:

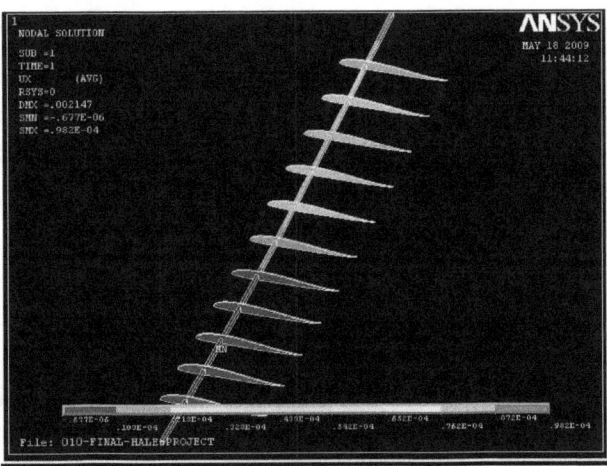

[Fig 16.22: Deflection in X-Direction]

Path Plot of Deflection in X-Direction:

A path has been defined on the main spar analysis results has been plotted on the graph as shown below;

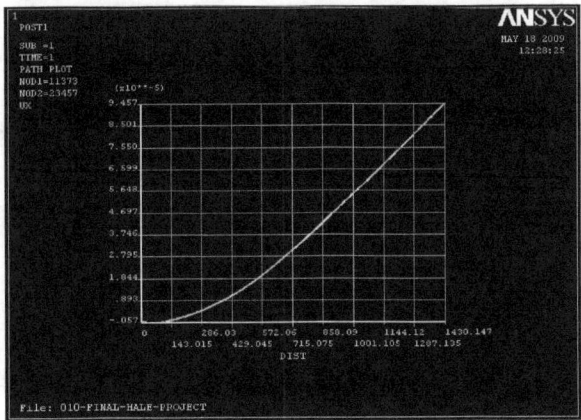

[Fig 16.23: Deflection Plot in X-Direction]

Deflection in Y-Direction:

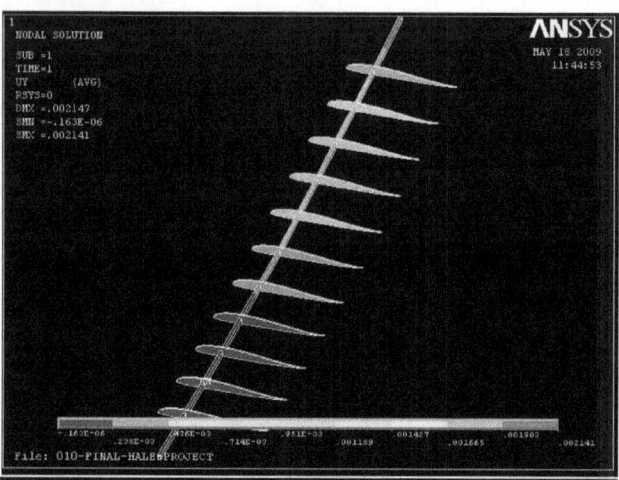

[Fig 16.24: Deflection in Y-Direction]

Graph of deflection in Y-Direction:

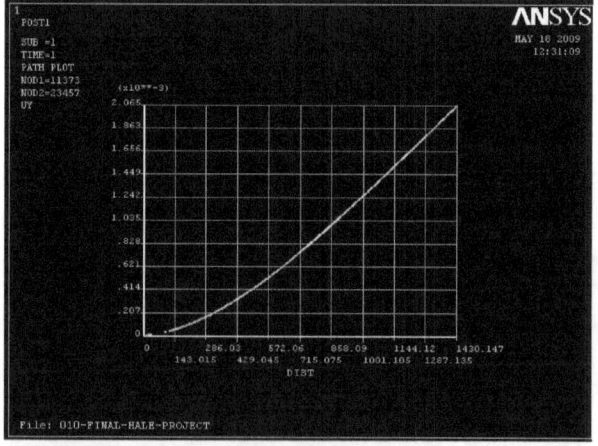

[Fig 16.25: Deflection Plot in Y-Direction]

Deflection in Z-Direction:

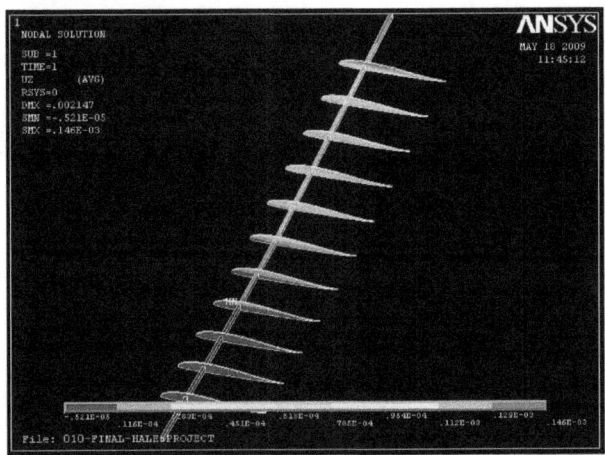

[Fig 16.26: Deflection in Z-Direction]

Graph of deflection in Z-Direction:

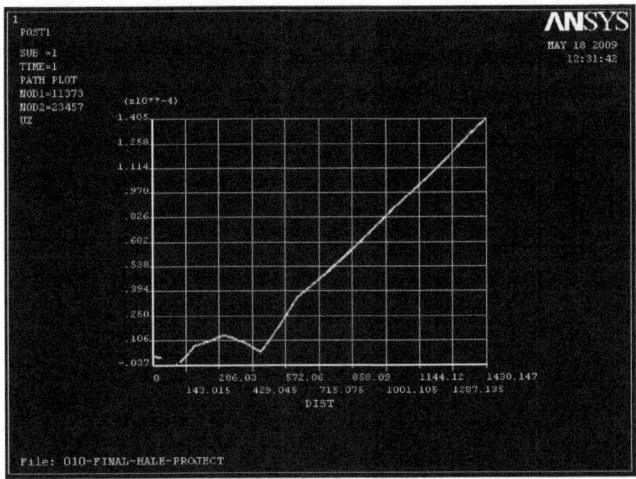

[Fig 16.27: Deflection Plot in Z-Direction]

Total Deflection:

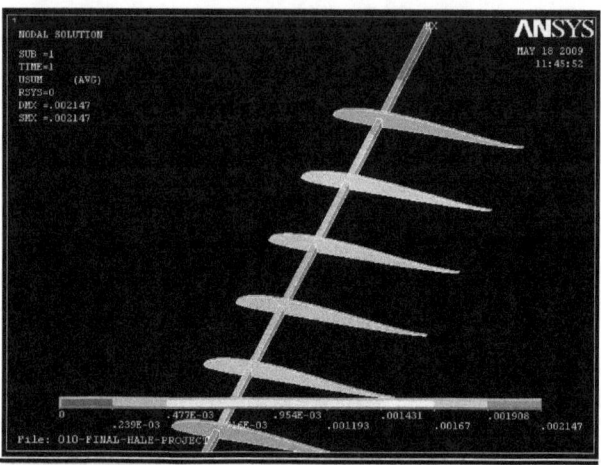

[Fig 16.28: Total Deflection]

Graph of total deflection:

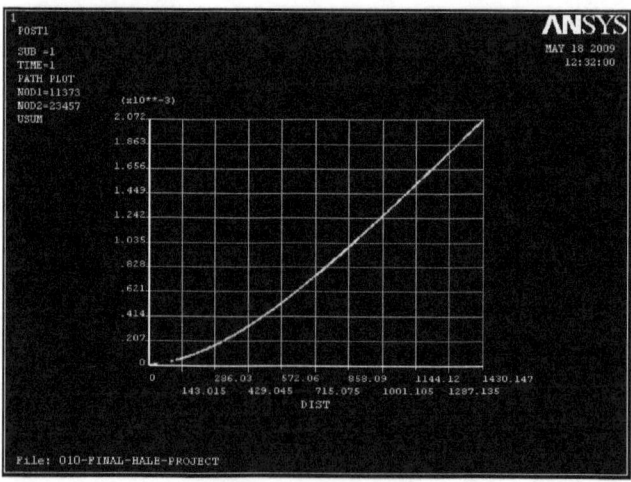

[Fig 16.29: Total Deflection Plot]

Graph of overall deflection:

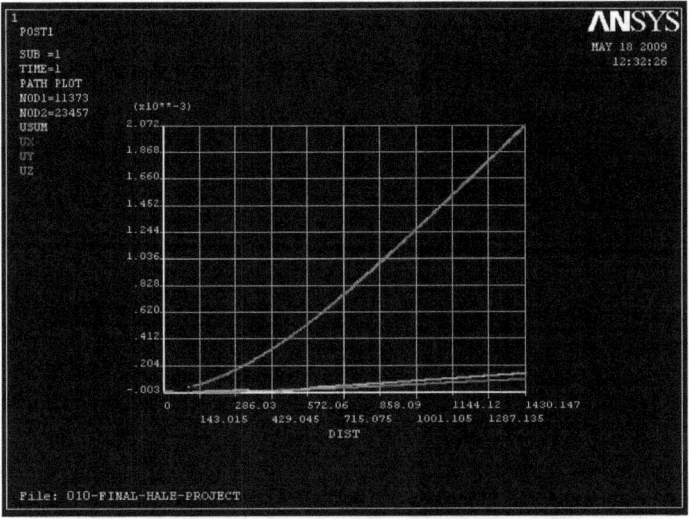

[Fig 16.30: Deflection Plot in XYZ]

16.4 Structural Analysis of Wing Panel:

After the optimization of wing main spar a structural analysis of a wing panel has been made in order to predict the structural loading behavior of wing.

16.4.1 Wing Panel Meshing:

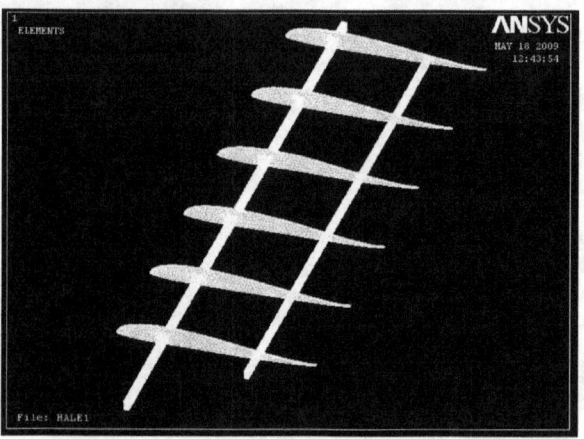

[Fig 16.31: Meshed Wing Panel]

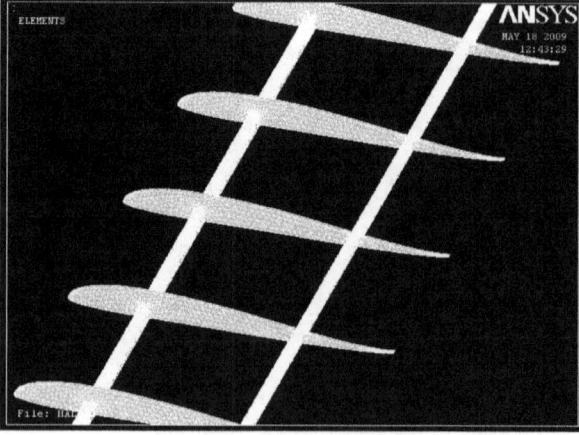

[Fig 16.32: Meshed Structure]

16.4.2 Loading on wing Spars:

Loads have been applied which are divided as 80% loads on the main leading edge spar and rest 20% load on the trailing edge spar of the wing.

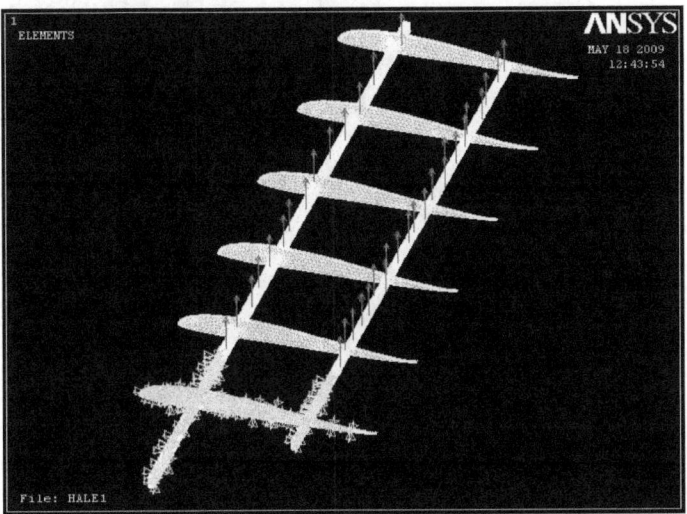

[Fig 16.33: Loading on wing spars]

16.4.3 Analysis Results:

Deformed Shape:

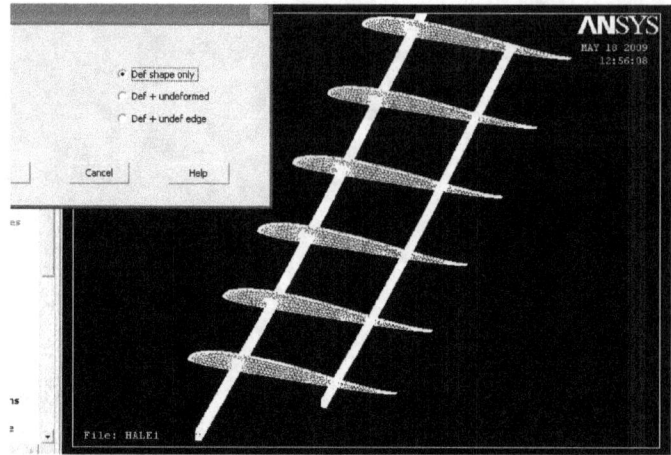

[Fig 16.34: Deformed Shape]

Deformation with undeformed Shape:

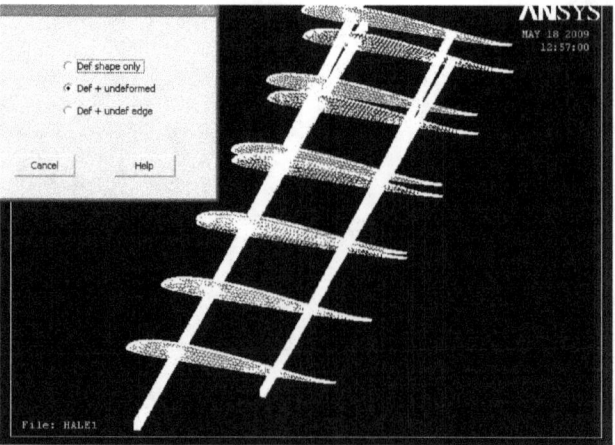

[Fig 16.35: Deformed & Un-deformed shape]

Deformation with undeformed Edges:

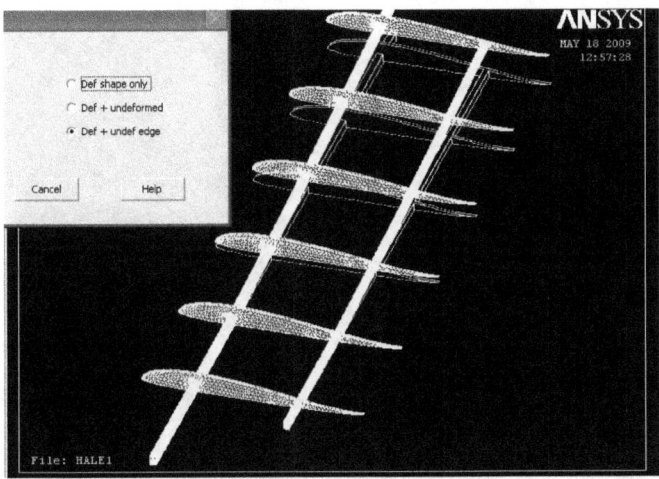

[Fig 16.36: Deformed & Un-deformed Edges]

Deflection in X-Direction:

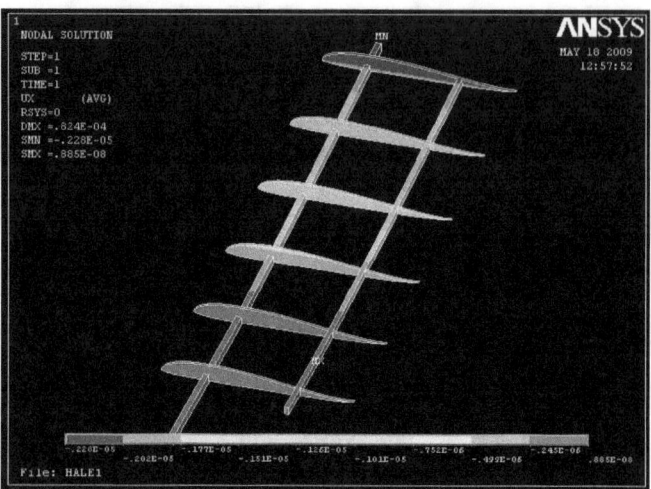

[Fig 16.37: Deflection in X-Direction]

Path Plot of Deflection in X-Direction in main Spar:

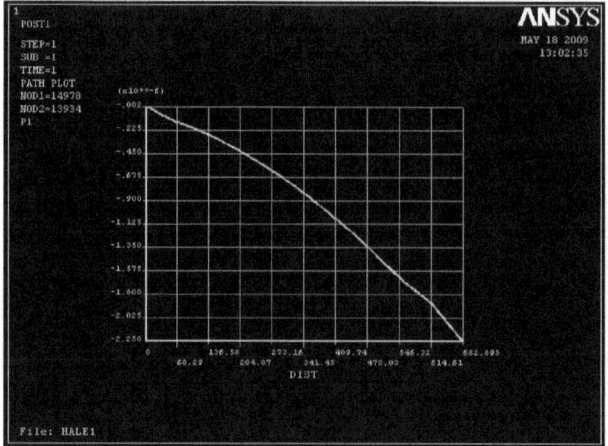

[Fig 16.38: Deflection Plot in X-Direction in Main Spar]

Path Plot of Deflection in X-Direction in Trailing edge Spar:

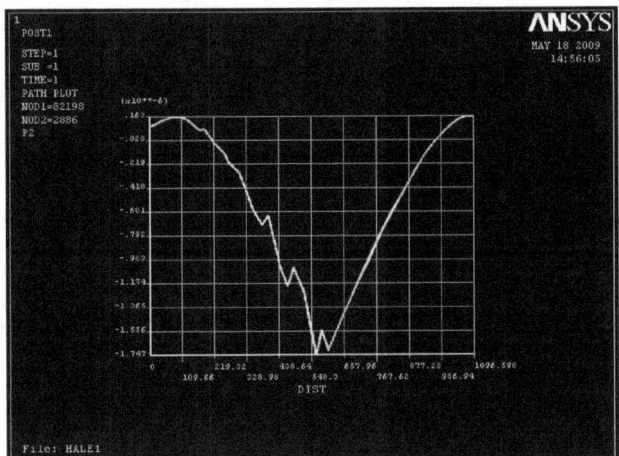

[Fig 16.39: Deflection Plot in X-Direction in Trailing Edge Spar]

Deflection in Y-Direction:

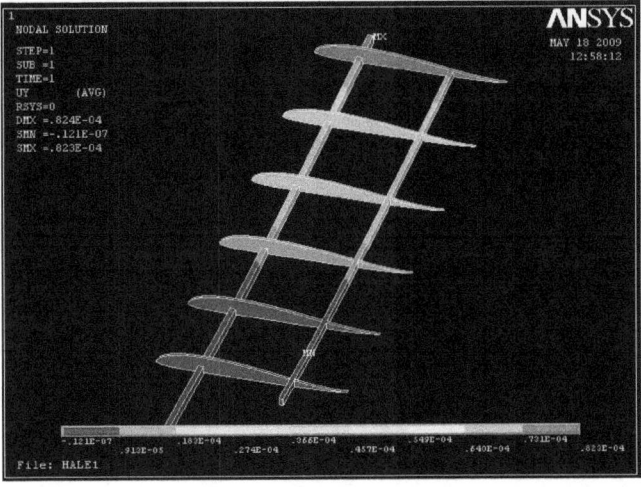

[Fig 16.40: Deflection in Y-Direction]

Path plot of deflection in X-Direction in main spar:

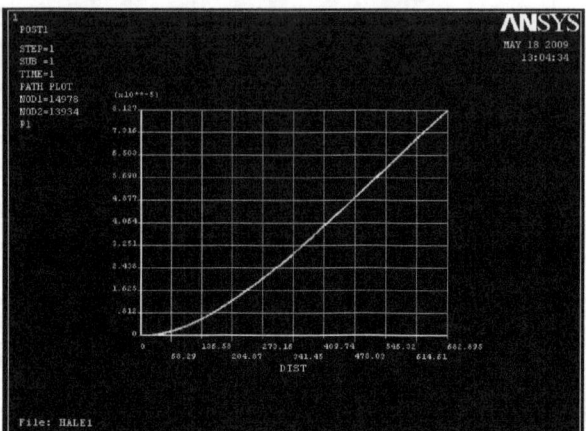

[Fig 16.41: Deflection Plot in Y-Direction in Main Spar]

Path Plot of Deflection in Y Direction in Trailing Edge Spar:

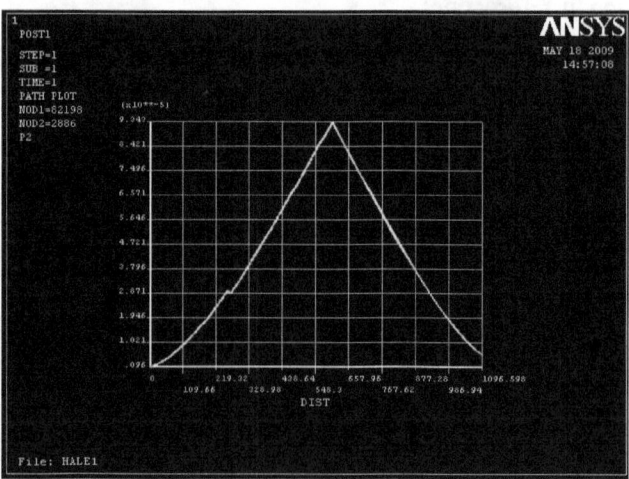

[Fig 16.42: Deflection Plot in Y-Direction in Trailing Edge Spar]

Deflection in Z-Direction:

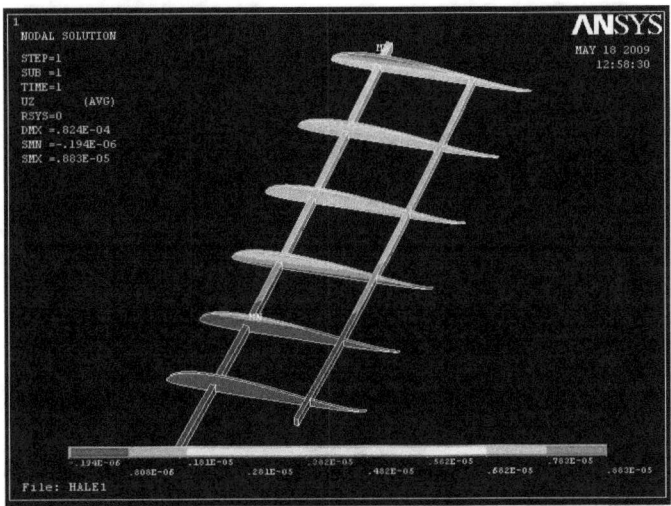

[Fig 16.43: Deflection in Z-Direction]

Path plot of deflection in Z-Direction in main spar:

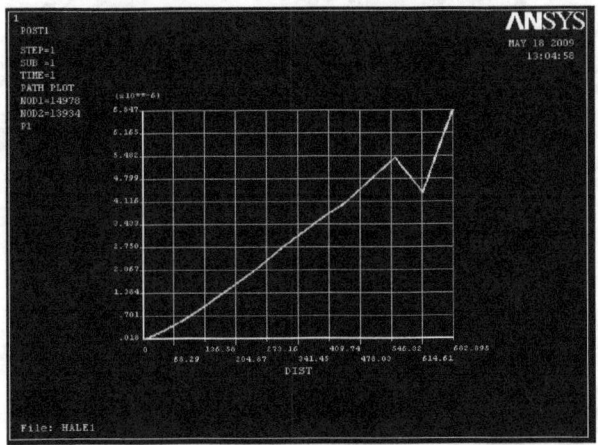

[Fig 16.44: Deflection Plot in Z-Direction in main spar]

Path Plot of Deflection in Z-Direction in Trailing Edge Spar:

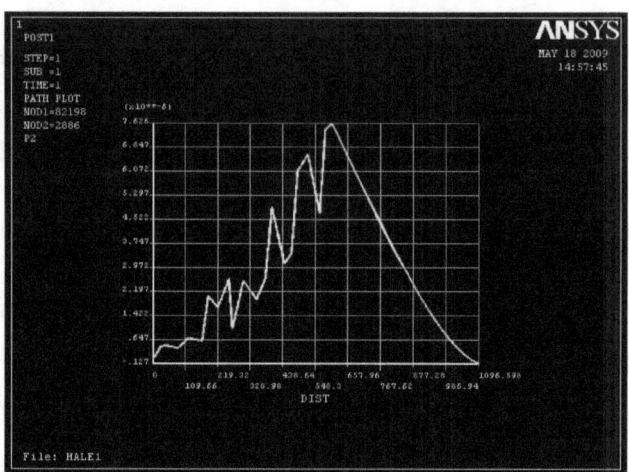

[Fig 16.45: Deflection Plot in Z-Direction in Trailing Edge Spar]

Total Deflection:

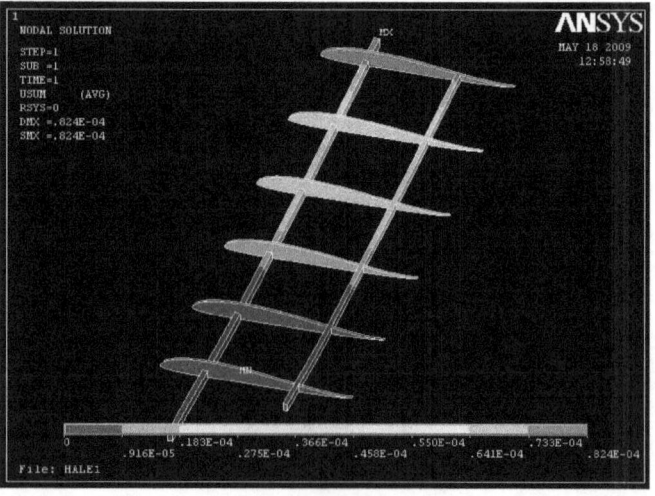

[Fig 16.46: Total Deflection]

Plot of total deflection in main spar:

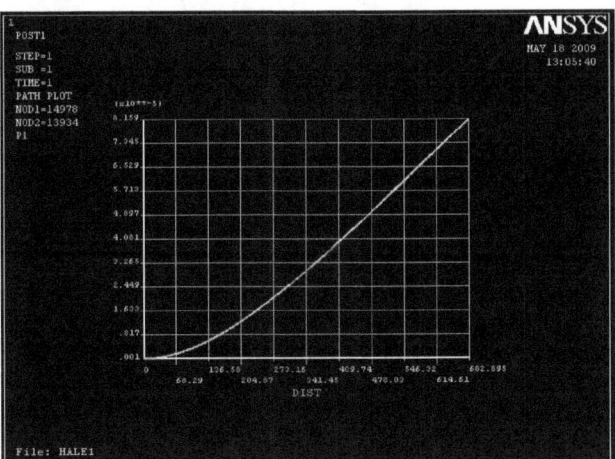

[Fig 16.47: Total Deflection Plot in main spar]

Path Plot of total deflection in trailing edge spar:

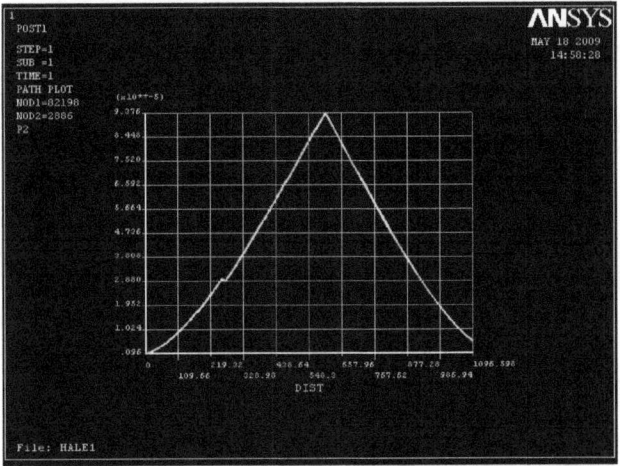

[Fig 16.48: Total Deflection in Trailing Edge Spar]

16.5 Conclusion:

From structural analysis of different spars configuration it has been concluded after the comparison of structural analysis result that placing spar at 20% of chord will result in less in less deformation and thus more favorable position. Further, wing panel structure analysis results in deformation in the tips of the wing panel and at the points of joints of the wing. Load is distributed between the main and trailing edge spar in the ratio of 80% and 20% of the total load. Trailing edge spar has been placed at 75% of the chord and structure analysis shows that trailing edge spar has shown more deformation or deflection in the structural configuration. Moreover, deformation has not a uniform trend in the wing panel.

Chapter 17 | Fluid Analysis

17.0 Fluid Analysis

The flow analysis of SAIST is done using FLUENT at steady level flight conditions. Fluid analysis involves the following processes:

17.1 Flow Analysis Steps

The steps involved are as:

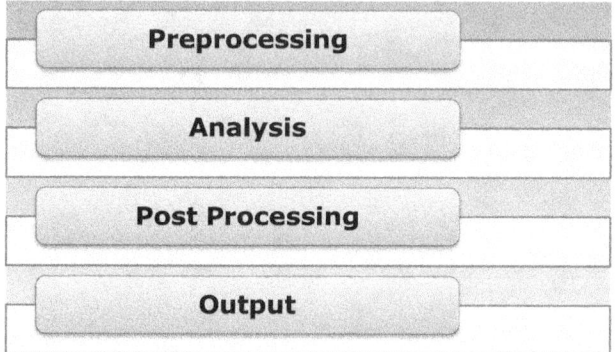

[Fig 17.1: Flow Analysis Sequence]

17.1.1 Preprocessing

Preprocessing involves,

- Geometry modeling
- Grid Generation
- Transforming into FLUENT
- Defining Flight Conditions
- Defining Boundary Conditions

17.1.2 Analysis

This phase of the analysis involves the simulation of the meshed geometry at desired numbers of iterations in order to converge on the appropriate solution.

17.1.3 Post Processing

Post processing doesn't involve any particular steps. It is basically verification of output and results according to prediction or estimation.

17.1.4 Output

Finally, output of the contours is obtained of different flow characteristics as pressure, temperature and velocity.

17.2 Grid Generation

Grids have been generated into Gridgen as shown below;

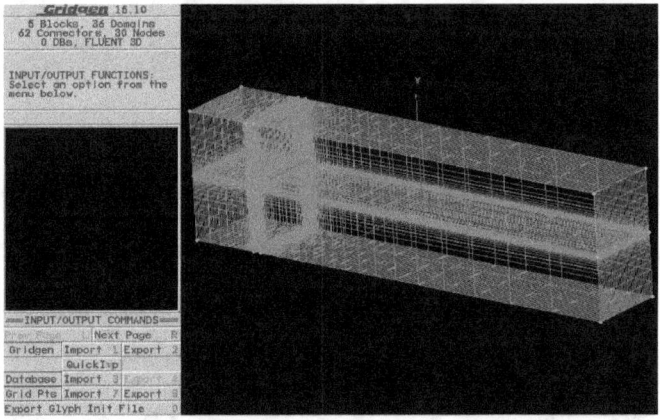

[Fig 17.2: Grid generation]

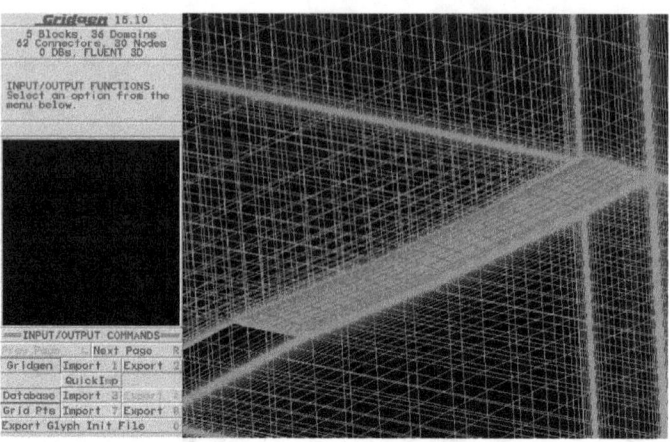

[Fig 17.3: Structured Grids along Geometry]

17.2.1 Grids in FLUENT

Grids generated are transferred into FLUENT for further analysis.

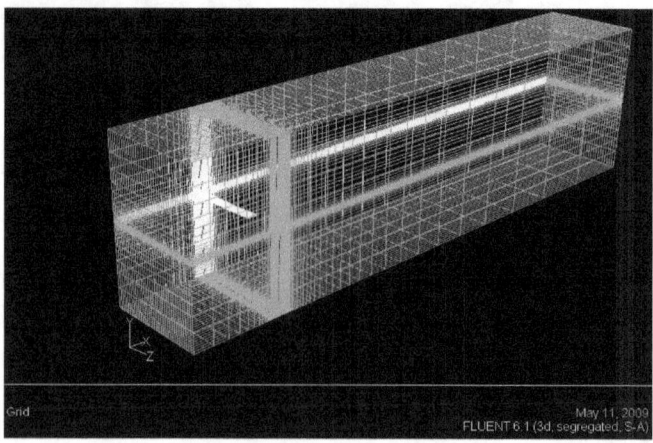

[Fig 17.4: Grids display after transfer in FLUENT]

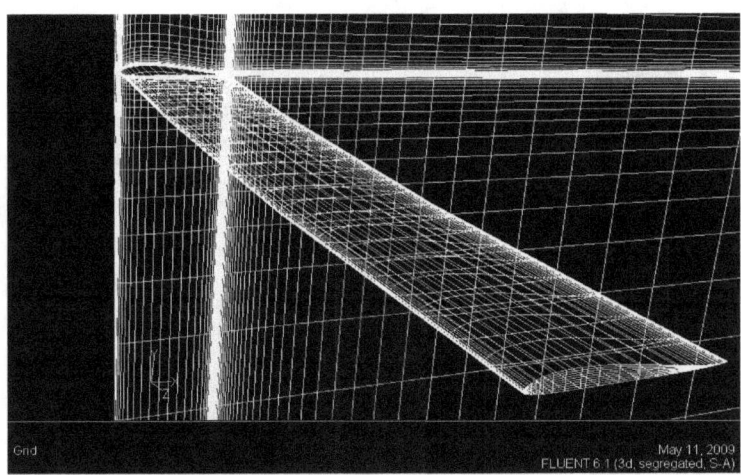

[Fig 17.5: Structured Grids along SAIST HALE]

17.3 Analysis Results

Analysis has been done on 60,000ft altitude flight conditions and at 74ft/sec.

17.3.1 Contours of Static Pressure

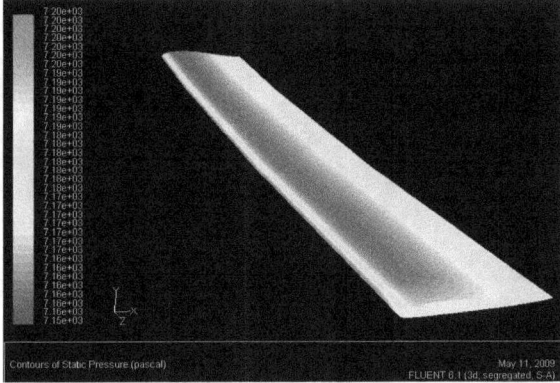

[Fig 17.6: Contours of Static Pressure]

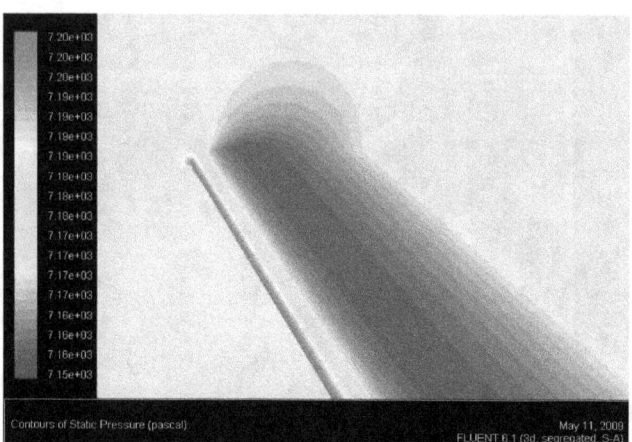

[Fig 17.7: Contours of Static Pressure across flow field]

17.3.2 Contours of Dynamic pressure

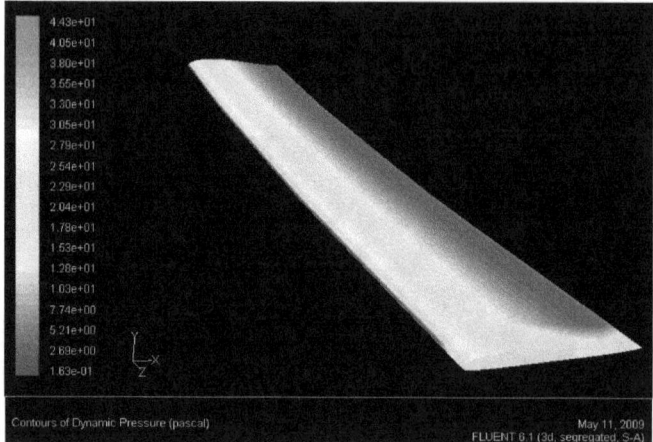

[Fig 17.8: Contours of Dynamic Pressure]

[Fig 17.9: Contours of Dynamic Pressure in flow field]

17.3.3 Contours of Total Pressure

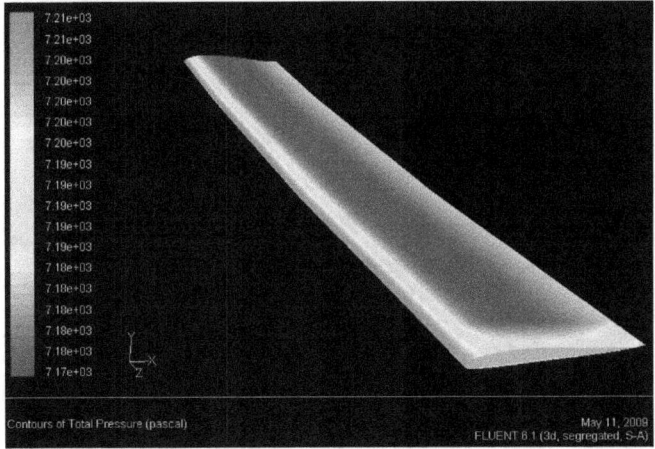

[Fig 17.10: Contours of Total Pressure]

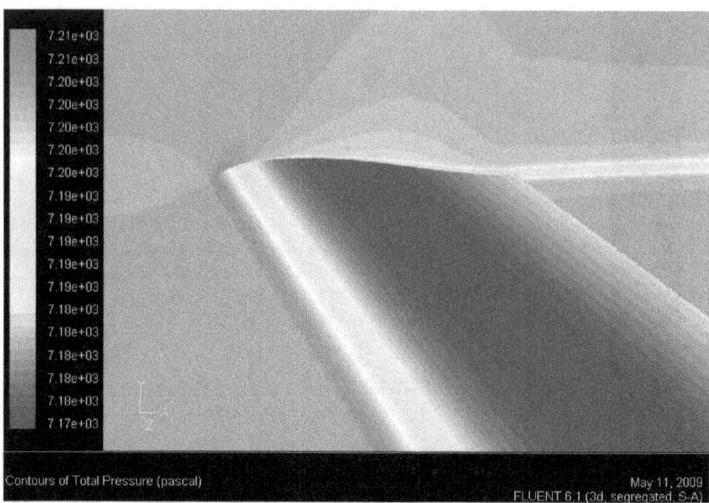

[Fig 17.11: Contours of Total Pressure in flow field]

17.3.4 Contours of Mach number

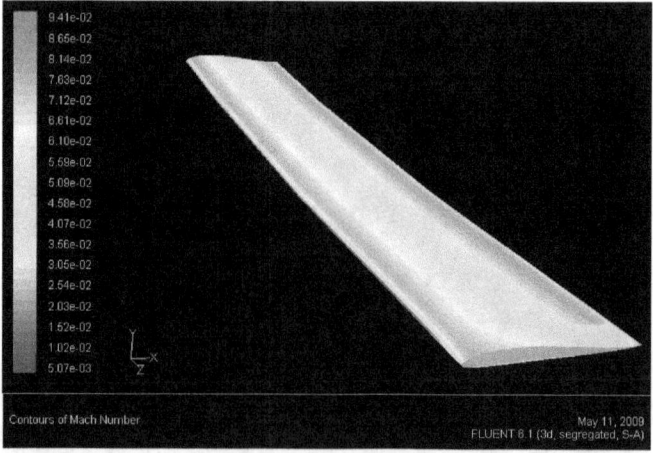

[Fig 17.12: Contours of Mach number]

[Fig 17.13: Contours of Mach number in flow field]

17.3.5 Contours of Velocity

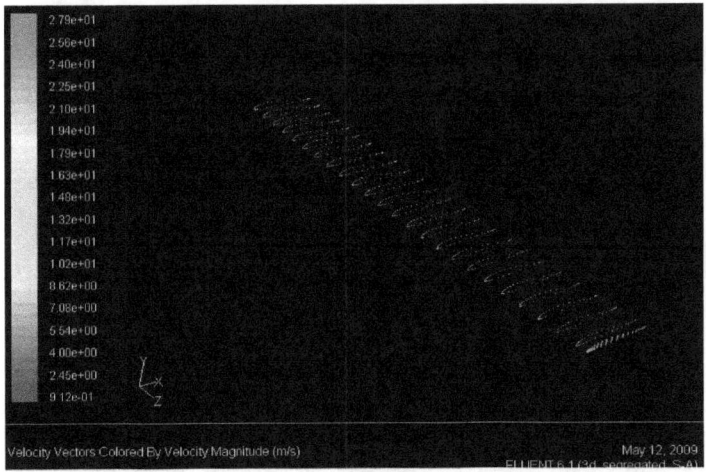

[Fig 17.14: Velocity Vectors colored by Velocity magnitude]

[Fig 17.15: Velocity Vectors colored by Velocity magnitude along far field]

17.3.6 Contours of Total Temperature

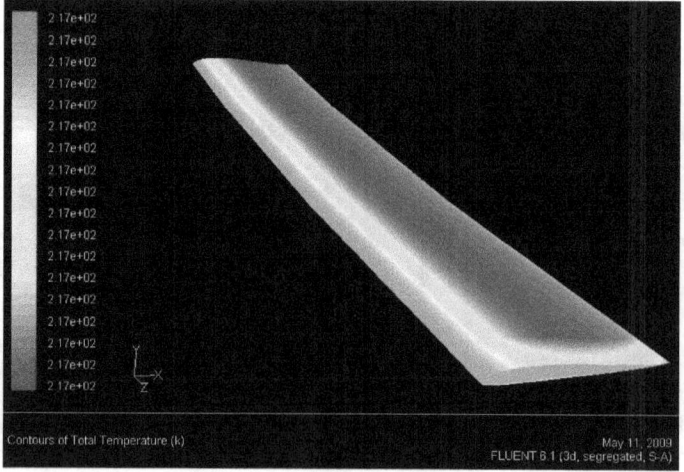

[Fig 17.16: Contours of total temperature]

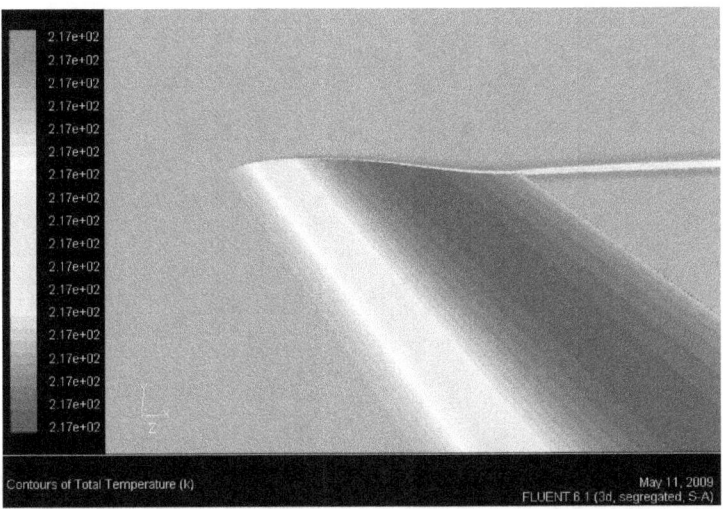

[Fig 17.17: Contours of Total Temperature]

17.4 Conclusion

Fluid analysis has been done in order to observe the pressure variation, Mach number variation, temperature variation and velocity profile along chord wise of the wing. Analysis gives us appropriate results in comparison with the theoretical estimations.

Chapter 18 | Detail Drawings

18.0 Detail Drawing

Drawings of the SAIST were made in AutoCAD 2008 Software to clarify the locations of the basic components of the SAIST. The components that were drawn in software were motors, motor mounts with motors, spars, ribs, location of the payload room, solar mechanism and autopilot and navigation equipments, and the external sheeting of the SAIST. These components were given different colors so that they could be distinguished. The drawing was made on full scale.

18.1 Major Components of SAIST

18.1.1 Motor with Cowling

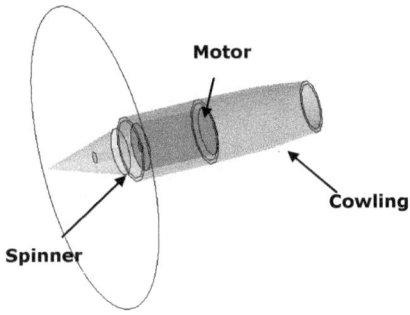

[Fig 18.1: Drawing of Motor with Cowling and Spinner]

18.1.2 Wing Rib

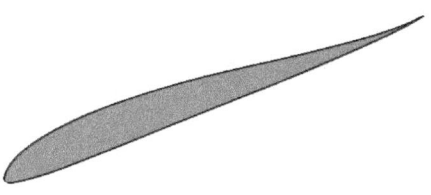

[Fig 18.2: Rib]

18.1.3 Spars

There are two spars being used in the structure of SAIST. One is the Leading edge spar which is an I-shaped Beam and the other is the Trailing edge spar whose shape is determined according to the shape of the airfoil.

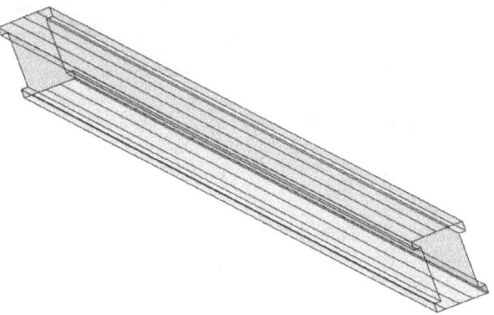

[Fig 18.3: Leading Edge Spar]

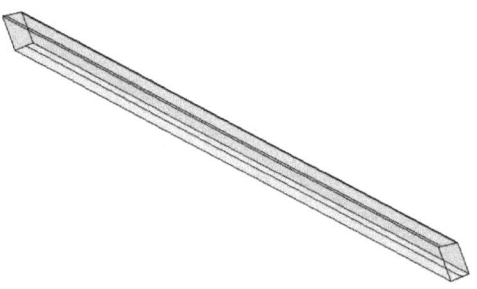

[Fig 18.4: Trailing Edge Spar]

18.2 Assembly Drawing

Following is the assembly drawing of SAIST showing the components in unassembled form.

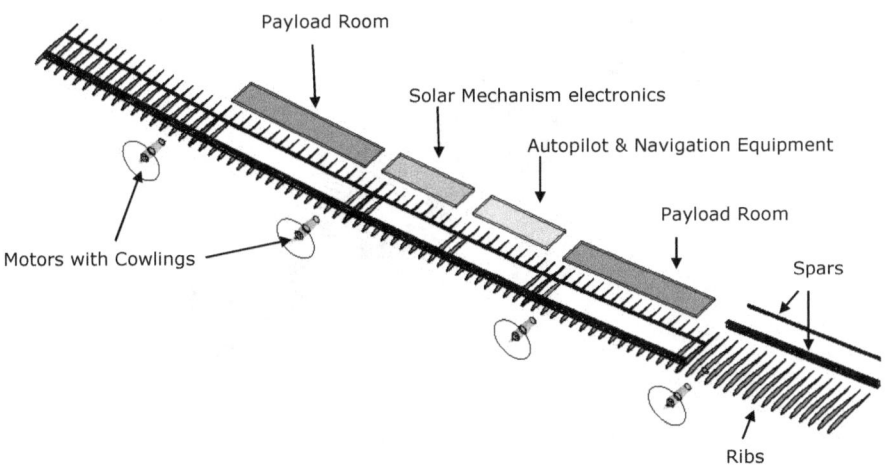

[Fig 18.5: Assembly Drawing showing various components of SAIST unassembled]

18.2.1 Payload Room

The Payload room shown in the above figure is the space in the SAIST for any equipment depending upon the mission. It may include communication equipment or reconnaissance equipment.

18.2.2 Solar Mechanism Electronics

The solar mechanism electronics box shown in the above figure consists of all the equipment required for the in-flight charging and continuous flight of the aircraft.

18.2.3 Autopilot and Navigation Equipments

The box shown in the figure may consist of the autopilot and navigation equipment for the autonomous flight of the aircraft.

18.2.4 Final Drawing

The final assembled drawings of SAIST are shown below.

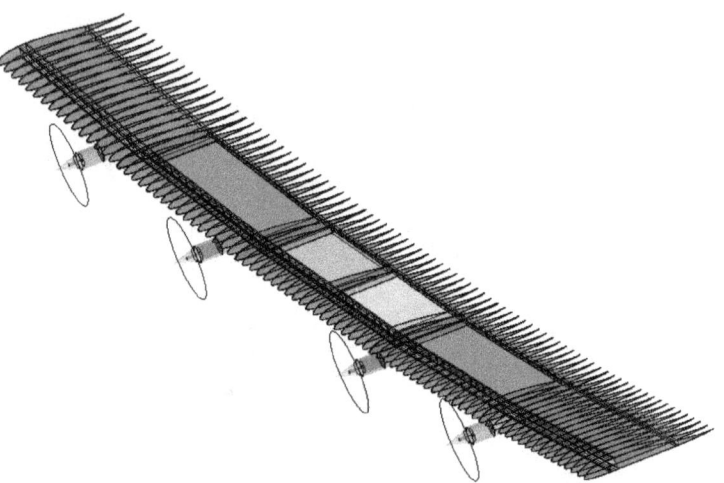

[Fig 18.6: Drawing without external sheeting]

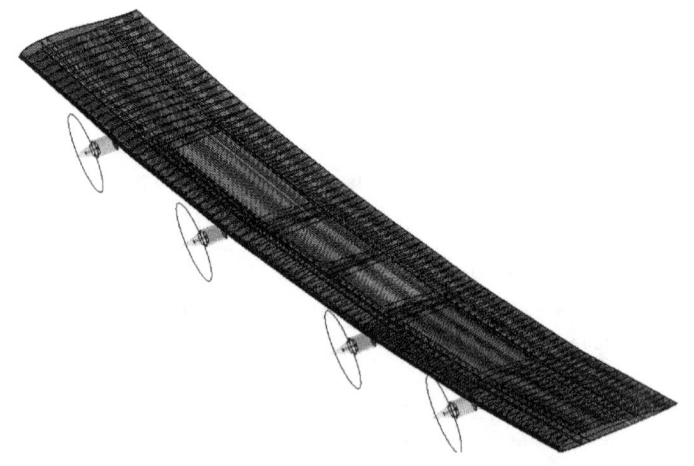

[Fig 18.7: Drawing with external sheeting]

Chapter 19 | Prototype Manufacturing

19.0 Prototype Building and Manufacturing

19.1 Introduction

In order to demonstrate the concept of SAIST, it was decided to build a scaled down model of the aircraft to exactly figure out the behavior that the configuration of the aircraft shows when it goes in to the air. The aircraft would still be having solar panels, but running the receiver's battery only, having two brushless motors with their respective compatible battery packs and speed controllers. The reason for using solar panels to charge only the receiver's battery was just to demonstrate the concept of in-flight charging in the SAIST. Furthermore, the surface area of the wing was not enough to have solar panels running the motors. Moreover, the scaled down model would be having landing gears in its configuration. The reason for using landing gears is to first ensure the stable flight of the aircraft and then the lading gears could be removed after having planned a proper take-off sequence of the aircraft.

This chapter would discuss the prototype building and manufacturing of the SAIST; explaining the sizing of the prototype and then the various techniques that were used for the construction of the SAIST.

19.2 Prototype building

Keeping in view the budget and construction constraints, it was decided to build a prototype of the SAIST that would ideally represent the concept at the same time finishes in time. Therefore, it was decided to build a 1/6th scale model of the SAIST. The design parameters were obtained using the same procedure as were obtained for the full scale model of SAIST. Following table shows some design parameters of the 1/6th scale prototype:

Span	12.5ft
Chord	1ft
Area	12.5ft^2
Aspect Ratio	12.5
Power loading	23.2
Thrust-weight ratio	0.201
Wing loading	1.44 lb/ ft^2
Lift-Drag ratio	17.13
Max. Velocity	70ft/s
Stall Velocity	37ft/s

[Table 19.1: Scale Model Characteristics]

The prototype would be having same number of panels as the full scale model has i.e. five. The division of the panels and their sizes were determined and shown in the following figure:

[Fig 19.1: Panels Division]

Side panel 1	2.417ft
Side panel 2	2.417ft
Mid panel	2.828ft
Side panel 3	2.417ft
Side panel 4	2.417ft

[Table 19.2: Panels Dimensions]

19.3 Manufacturing

The manufacturing phase includes selection of a material and then construction of the wing using various building techniques.

19.3.1 Material

The selection of material for the prototype construction involves several parameters which are to be analyzed before selecting the final material. The material to be selected should have less weight at the same time providing adequate strength. Keeping in view the construction facilities available, the skill level for the material should be selected accordingly. The material should be affordable too.

Keeping in view all of these factors, some figure of merits for the selection of material were generated for the selection of material for the aircraft.

19.3.2 Figure of merits

Following figure of merit were taken into account while selecting a material and eventually a manufacturing process for the construction of SAIST.

	Figure of Merit(FOM) for Material Selection		
S. No.	Figure of Merit	Weight	Remarks
1	Weight Saving (WS)	+1 to -1	+1 to -1 for lighter to heavier materials
2	Availability (AV)	+1 to -1	+1 for easily available and -1 for not available
3	Skill Level (SL)	+1 to -1	+1 to -1 for available construction facility
4	Affordability (AF)	+1 to -1	+1 to -1 for economic factor
5	Time Requirement (TR)	+1 to -1	+1 to -1 for quick and slow production rate
6	Strength (S)	+1 to -1	+1 to -1 for adequate strength or weak material

[Table 19.3: FOMs for Material Selection]

19.3.3 Selection of Material

The selection of the right material for the construction of SAIST was done by forming a decision matrix using the same FOMs generated previously. Weights were assigned to each FOM and final selection was made.

	Material Selection Matrix							
S. No.	Material	FOM						Total
		WS	AV	SL	AF	TR	S	
1	Composites	+1	0	-1	+1	-1	+1	+1
2	Balsa and Plywood	0	+1	+1	+1	+1	0	+4
3	Light Aluminum Alloys	-1	-1	0	+1	0	+1	0

[Table 19.4: Material selection matrix]

According to the decision matrix, balsa and Plywood get the maximum weight age. Therefore, Balsa and plywood were selected for the construction of SAIST.

19.3.4 Landing Gear Selection

A quadric-cycle landing gear configuration was chosen for SAIST. For the construction of such a configuration, the landing gears are needed to be strong enough to bear the load of the aircraft and the jerks it would face during the landing but they have got to be light in weight as the aircraft is intended to be a light weight structure. The material selection for the landing gears was done using decision matrix based on the same figure of merits which were used for the selection of material for the wing construction.

S. No.	Material	FOM						Total
		WS	AV	SL	AF	TR	S	
1	Composites	+1	-1	-1	+1	0	+1	+1
2	Carbon Steel rods	0	+1	+1	+1	+1	0	+4

[Table 19.5: Landing gear material selection]

19.3.5 Manufacturing phases

Following is the chart showing the different manufacturing phases for the construction of SAIST.

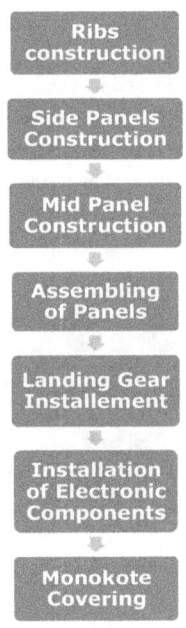

[Fig 19.2: Manufacturing Phases Sequence]

Each of these phases is briefly discussed below.

19.3.6 Ribs Construction

Construction of ribs was done using Balsa wood. An aluminum template of the rib was cut using cutter. This template was then used to construct the ribs by mounting the template over the balsa sheet and cutting it according to the aluminum rib profile.

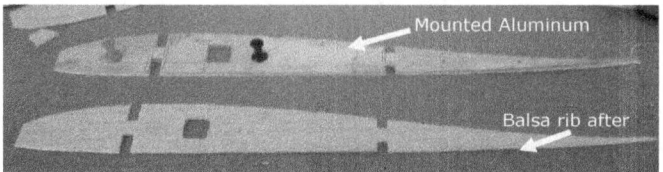

Mounted Aluminum

Balsa rib after

[Fig 19.3: Ribs Construction]

19.3.7 Side Panels Construction

After the construction of ribs, side panels were constructed using a plan made on AUTOCAD software. Both leading edge and trailing edge spars were used in the construction of the side panel. These spars were made from balsa wood sheets. Balsa sheet webbing was done to strengthen the spars. It was made sure that the contact lower surface contact points of all the ribs constituting the spars are same so that no twist is induced in the panel. Furthermore, the construction of the side panel was done on a leveled surface so that no warping is induced in it.

[Fig 19.4: Wing Panel Construction]

[Fig 19.5: Side Panel Construction]

19.3.8 Mid Panel Construction

The Mid Panel was constructed to be strong enough to bear the maximum loads because it is the mid panel to which the loads of the side panels get concentrated. Apart from that, the landing gears, motors, Speed controllers, batteries and receivers were to be mounted on the mid panel. For this reason, the spar of the mid panel was made using laminated plywood. Balsa webbing was done to further strengthen the spar. The Mid panel and side panels were sheeted from above and below using balsa sheet. This further strengthened the structure.

[Fig 19.6: Mid Panel Construction]

19.3.9 Assembling of Panels

After constructing four side panels and a mid panel, all of the five panels were assembled together. The inboard side panels had to be at a dihedral angle of 2 degrees with the mid panel and the outboard side panels at a dihedral angle of 4 degrees with the mid panel. For this purpose, plywood braces were cut for connecting the two panels at the required angle. it was made sure that the panels on both sides of the mid panel are at same angles.

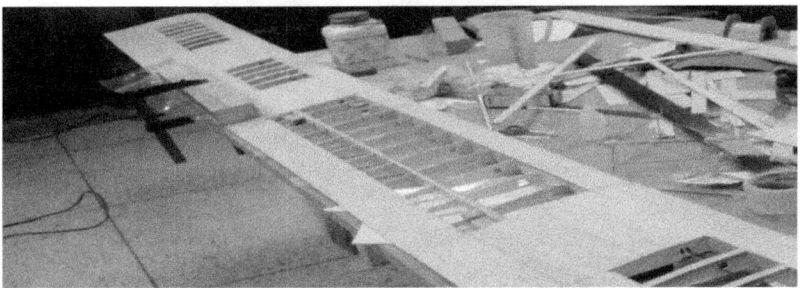

[Fig 19.7: Assembling of Panels]

A layer of fiber glass was used to further strengthen the joints between the two panels as shown below.

[Fig 19.8: Fiber Glass layer]

19.3.10 Landing Gear Installment

Landing gears were constructed using carbon steel rods. The rods were molded in to the required shape by heating the rods at required locations. Bass wood parts were installed in the mid panel to support the landing gear steel rods. The rods were attached with the bass wood using epoxy. It was made sure that there is enough clearance between the landing gear rods and the propeller to be mounted on the mid panel so that the propeller may not hit the gear during its operation. The landing gears were then sandwiched between two sheets of balsa giving it an aerodynamic shape.

[Fig 19.9: Landing gears installment]

19.3.11 Installation of Electric Components

The electric components used in SAIST were servos, receiver, speed controllers, receiver and motor batteries, and motors. One servo had to be installed in each of the side panels for control surface operation. For that purpose, servo boxes were made for the installation of servos as shown below.

[Fig 19.10: Servo Box]

For installation of motors, motor mounts were constructed in the mid panel and plywood was used to strengthen the structure.

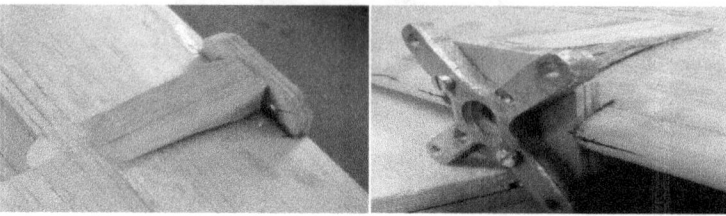

[Fig 19.11: Motor mounts construction]

[Fig 19.12: Installed Motor]

Boxes for receivers and speed controllers were also constructed in the mid panel for their installation and the boxes were covered with plywood lids to protect the components as shown below.

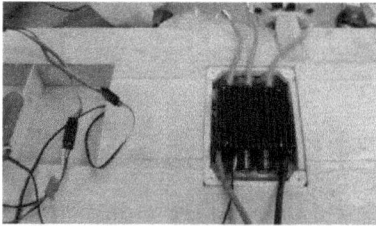

[Fig 19.13: Speed controller and receiver boxes]

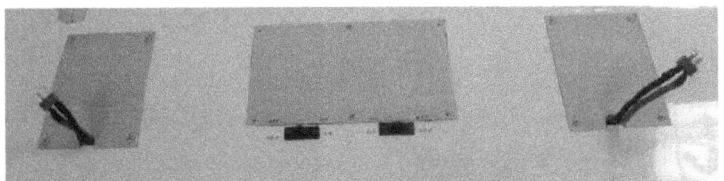

[Fig 19.14: Prepared Boxes]

19.3.12 Aircraft Assembly

After fabricating all components and panels SAIST is assembled finally as shown in the figure below:

[Fig 19.15: Aircraft Assembly]

19.3.13 Monokote Covering

After finagling the electronic and structural components, the aircraft was sanded to smooth the surface. After sanding, the aircraft was covered with thin monokote covering using heat sealing iron. The finalized aircraft after covering is shown below.

[Fig 19.16: Final Aircraft]

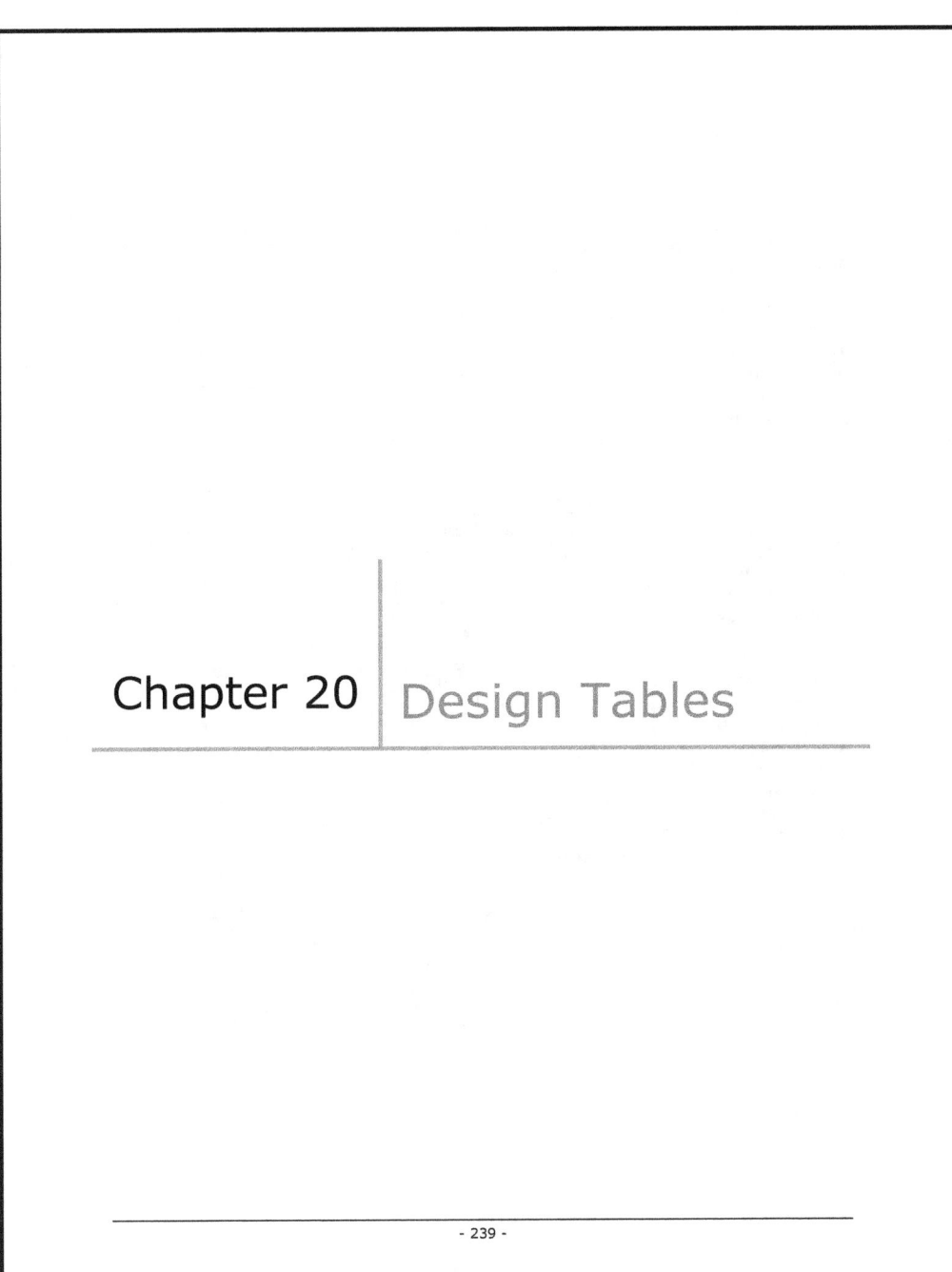

Chapter 20 | Design Tables

20.0 Design Tables

SAIST design tables and design summary is as:

Geometry	
Airfoil	mhmi2
Span (ft)	86ft
Chord (ft)	6.139ft
Area (ft^2)	528ft^2
Aspect Ratio	14
Inboard Dihedral (°)	2°
Tip Dihedral (°)	4°
Twist Distribution	0°

[Table 21.1: SAIST Geometry]

Performance	
$(L/D)_{max}$	14.226
Cruise velocity (ft/s)	74
Wing Loading (lb/ft^2)	0.612
$(ROC)_{max}$ (ft/s)	14
P_A (hp)	6.37
P_R (hp) (At Cruise)	3.09
Takeoff Distance (ft)	830.36
Landing Distance (ft)	900
Turning Distance (ft)	328.65
Accelerated Climb Distance (ft)	173.51
Maximum Load Factor 'n'	3.44
Roll angle (Rad)	1.27
Turn Rate (Rad/s)	1.43
Glide Path angle (°)	3.21

[Table 21.2: SAIST Performance]

Propulsion	
Motor	Hacker Brushless A60-18M
Propeller	APC-E 34×10
Battery	LiPo 10 cell
Speed controller	Brushless ESC- 100 Amp

[Table 21.4: SAIST Propulsion]

Aerodynamics	
Airfoil Lift Curve Slope (rad^{-1})	6.20275
Aircraft Lift Curve Slope (rad^{-1})	5.32695
Induced Drag Coefficient	0.03611
Total Skin Friction Drag	0.00930
Induced Drag	0.0361
Parasite Drag	0.0186
Form Drag	0.0165
Trim Drag	0
Cooling Drag	0
Interference Drag	0
Skin Friction Drag	0.0093
Wind milling drag	0.00010793
Stopped propeller drag	0.00001212
Landing Gear Drag Coefficient	0.9833
Total Zero Lift Drag Coefficient	0.03529

[Table 21.3: SAIST Aerodynamics]

Stability	
X_{ac} (ft)	1.6
X_{cg} (ft)	1.16
S.M (%)	7
C_{mo}	0.0151
C_{mw}	0.0132
Downwash angle derivative	0.25
Upwash angle derivative	1.95

[Table 21.5: SAIST Stability]

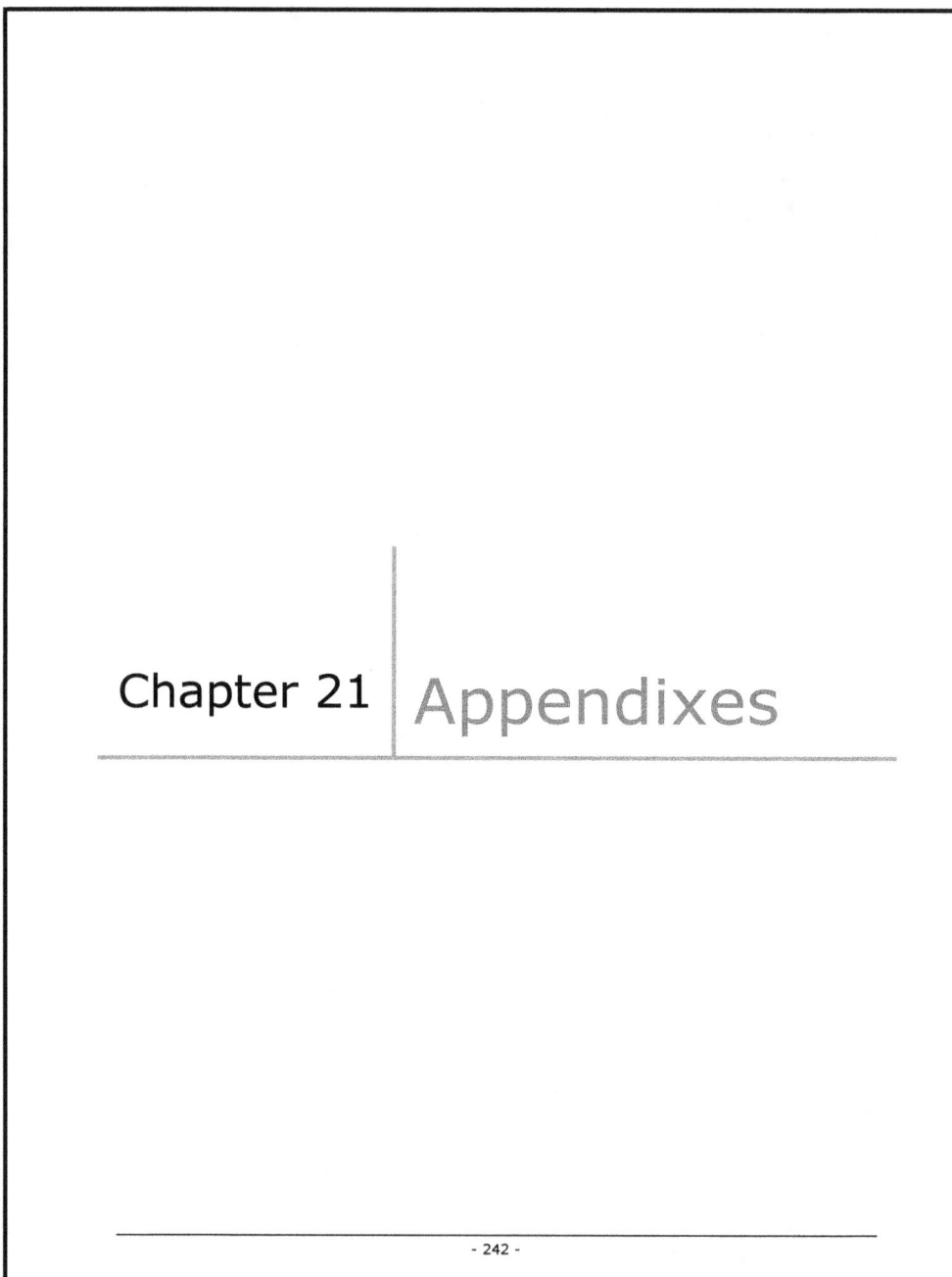

Chapter 21 | Appendixes

21.0 Appendixes

21.1 Appendix A

Sailplane Data:

Sailplane data has been evaluated for comparison with the preliminary sizing of SAIST HALE UAV. So, for reference attached hereby as;

Aircraft Name	Span (ft)	AR	Wing Area (ft-sq)	Weight (lbs)
Alexander Schleicher	59.2	15.2	230	1234
Schleicher Rhohbussard	47	14.5	151.8	540
Schleicher ka-1	32.8	10.1	106.6	430
Schleicher Ka-2	52.5	13.4	180.8	1058
Schleicher Ka-3	32.8	10.1	106.6	430
Schleicher Ka-4	42.6	10.3	176	900
Schleicher Ka-6CR	49.2	18.1	134	670
Schleicher Ka-6E	49.2	18.1	133.6	663
Schleicher Ka-7	52.5	14.6	189	1056
Schleicher Ka-8	49.2	15.9	153	683
Schleicher Ka-9	39.37	12	129.16	507.06
Schleicher Ka-10	49.21	17.96	134.87	705.47
Schleicher ASW-12	60	26	140	948
Schleicher ASK-13	52.5	14.6	188.4	1057
Schleicher ASK-14	46.9	16.8	136.5	794
Schleicher ASW-15	49.2	20.45	118.2	900
Schleicher ASW-17	65.6	27	160	1257
Schleicher ASK-18	52.49	19.71	139.8	738.54
Schleicher ASW-19	49.2	20	118	994
Schleicher ASW-21	49.2	16.1	193.21	794
Schleicher ASW-22	55.77	38.3	175.6	1653
Schleicher ASK-23	82	17.4	138.8	794
Schleicher ASW-24	49.2	22.5	107.6	1102
Schleicher ASH-25	49.2	38.3	175.6	1653
Schleicher ASH-26E	82	27.69	125.94	1289
Schleicher ASW-27	59	25	96.88	1102
Schleicher ASW-28	49.2	21.43	113	1202
Schempp-Hirth	49.2	21.3	113.9	1157

Schempp-ASW-20	49.2	21.4	113	1157
Schempp-Hirth-duo-discuss	65.2	24.4	24.4	1453
Schempp-Hirth-janus	65.6	23	23	1154
Schempp-Hirth-Mini NImbus	49.2	23	23	992
Schempp-Hirth-NImbus-2	66.6	28.6	28.6	1433
Schempp-Hirth-Nimbus-3	80.3	35.9	35.9	1546
Schempp-Hirth-Nimbus-3D	80.7	36	36	1764
Schempp-Hirth-Nimbus-4	86.6	38.8	192.2	1653
Schempp-Hirth-Nimbus-4D	86.9	39.1	193.32	1653
Schempp -Hirth-SHK	55.8	20.2	158	816
Schempp -Hirth-Standard Cirrus	49.2	16.7	146	772
Schempp -Hirth-Cirrus	49.2	22.5	107.5	860
Schempp -Hirth-Ventus	49.2	23.7	102.4	1157
Schempp -Hirth-Ventus-2	49.2	23.3	104.41	1157

Reference Plots from statistical data of sailplanes:

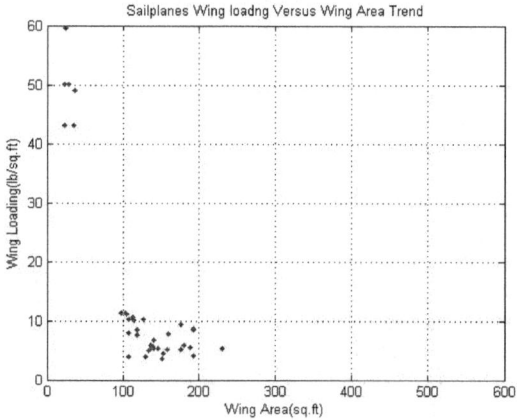

[Fig: Wing loading Versus Wing area trend distribution Plot]

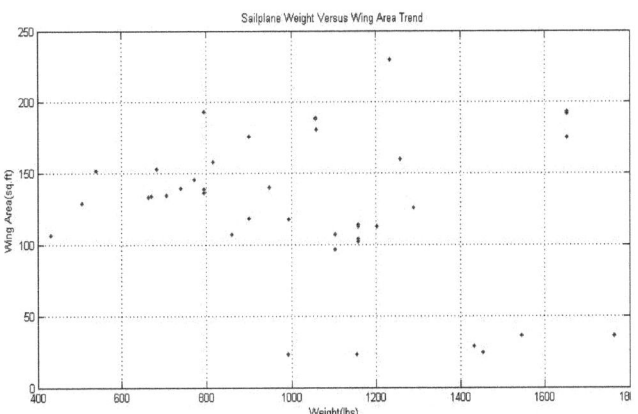

[Fig: Wing area versus weight trend distribution Plot]

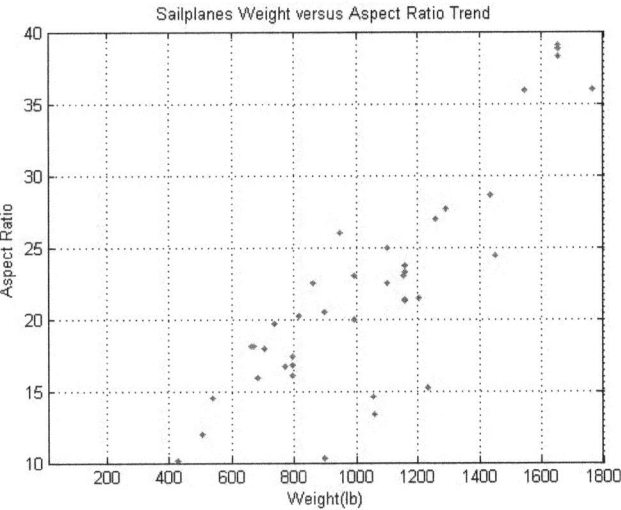

[Fig: Aspect ratio versus weight plot]

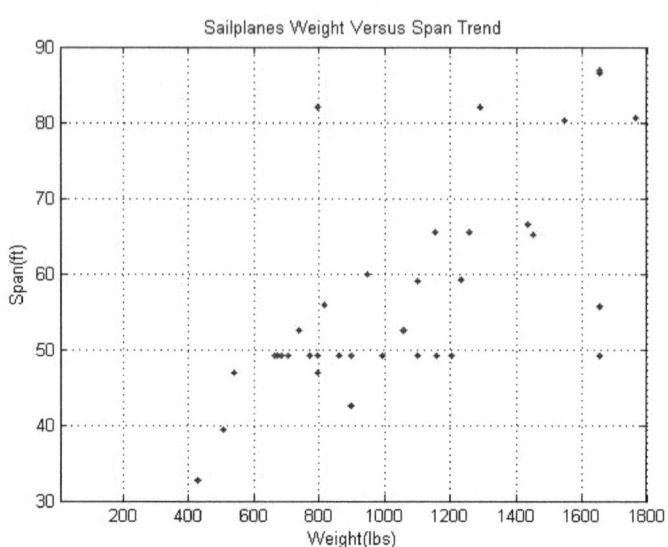

[Fig: Span versus Weight Distribution Trend]

21.2 Appendix B

Performance Plots Code

For reference;

```
%-----------------SAIST HALE UAV Performance-----------------%
%clc
%clear all
%v1=[0:1:200];
%v=v1.*1089.239;
%re=[0:1000:60000000];
%alt=[0 10000 20000 30000 40000 50000 60000 70000 80000];          %Altitudes
%den=[0.0023 0.0017 0.0012 0.00089 0.000587 0.00036 0.00022 0.00013 0.00086]; %Densities
%ve=[40 46.52639984 55.37749242 64.30265516 79.17806039 101.1050059 129.3339581
168.2488999 65.41459904];%Vel
%te= den.*0.5.*ve.^2;
%m=[0:0.01:1];a=1089.23;          %Mach no & Speed of sound
%w=323.13;                    %Weight
%s=528;                       %Wing Area
%ws=0.612;                    %Wing loading
%tw=0.242;                    %Thrust to weight ratio
%pw=0.05682;                  %Power to weight ratio
%ar=14;                       %Aspect ratio
%e=0.65;                      %Span efficiency factor
%nn=96.2;                     %Propeller RPS
%dd=1.75;                     %Propeller diameter
%for i=1:500
%q=(0.00022.*0.5.*(v).^2);          %Dynamic pressure
%k=0.03499;                   %Drag due to lift k factor
%cla=(w./(q.*s));             %Coefficient of lift
%cdo=0.0352910;               %Zero lift drag
%cdi=k.*(cla.^2);             %Induced drag
%cd=cdo+cdi;                  %Total drag coefficient
%d=q.*s.*cd;                  %Total drag-----Total thrust required
%p1=(q.*cdo)./ws;
%cfe1=(1.92)./((re).^0.5);          %Laminar skin friction drag
%cfe2=(0.63)./(log(0.056.*re)).^2;   %Turbulent skin friction drag
%cfe=cfe1+cfe2;               %total skin friction drag
%p2=ws./(3.14.*q.*ar*e);
%p=p1+p2;
%ld=1./p;                     %Lift to drag ratio
%sc=tw-(1./ld);
%ag=asin(sc);                 %Climb angle in radians
%tr1=q.*s*cdo;                %Parasite thrust required
%tr2=q.*s*k.*cla.^2;          %Induced thrust required
%pr=(v.*d)./0.42;             %Power required
%c1=cla.^(0.5);
%c2=cla.^(1.5);
%cl1=c1./cd;                  %CL.^1/2/CD
%cl2=c2./cd;                  %CL.^3/2/CD
%cl3=cla./cd;                 %CL/CD plot
%p=(6067.001-pr)./w;          %Rate of climb
%rc=v.*(0.242-(0.00001283.*v.^2)-(194.671./v.^2));%Rate of climb
```

```
%n1=cl3.*0.242;                          %Load factor in radians
%n=n1.*57.7;                             %Load factor in degrees
%phi=acos(1./n);                         %Roll angle/Bank Angle
%ra=(32.2.*((n1.^2-1).^0.5))./v;         %Turn radius
%ra1=(32.2.*(n1-1))./v;                  %Pull up maneuver
%ra2=(32.2.*(n1+1))./v;                  %Pull down maneuver
%pef=0.85./(1+(w./((cl3.*q).*(1.66^2)))); %Propeller Efficiency
%ct=(d./(0.00022*((nn^2)*(dd^4))));      %Propeller thrust coefficient
%adv=v./(nn.*dd);                        %Advance ratio
%cp=pr./(0.00022*((nn^3).*(dd^5)));      %Propeller power coefficient
%pe=adv.*(ct./cp);                       %Propeller efficiency
%csp=v.*((0.00022./(pr.*(nn^2))).^(1/5)); %Propeller speed power coefficient
%efe1=(((1+ct).^0.5)-1)./(((1+ct).^0.5)+1); %Propeller efficiency reduction-slipstream
%efe=efe1.*100;
%end
%plot(adv,pe)
%hold on
%plot(ve,alt,'red')
%hold on
%plot(re,cfe,'green')
%plot(cdi,cla)
%plot(v,p,'red')
%hold on
%plot(v,cl2,'blue')
%hold on
%plot(v,cl3,'green')
```

21.3 Appendix C

Datcom Stability Analysis

INPUT and OUTPUT files are attached here for reference:


```
****************************** INPUT DATA CARDS ******************************


 BUILD
 $FLTCON  NMACH=1.0, MACH(1)=0.1,
 NALPHA=9.0, ALSCHD(1)=-2.0,0.0,2.0,4.0,8.0,12.0,16.0,20.0,24.0,
 RNNUB(1)=5.622E4$
 $OPTINS SREF=528.0, CBARR=6.139, BLREF=86.0$
 $SYNTHS XCG=1.16641, ZCG=0.0, XW=0.0, ZW=0.0, ALIW=0.0, XH=0.0,
 ZH=0.0, ALIH=0.0, XV=0.0, VERTUP=.TRUE.$
 $WGPLNF CHRDTP=6.139, SSPNE=43.0, SSPN=43.0, CHRDR=6.139,
 SAVSI=0.0, CHSTAT=0.0, SWAFP=0.0, TWISTA=0.0, SSPNDD=33.3,
 DHDADI=2.0,DHDADO=3.0, TYPE=1.0$
 $WGSCHR TOVC=0.0931, DELTAY=0.00497, XOVC=0.0931, CLI=0.60, ALPHAI=4.80,
 CLALPA(1)=0.1075, CLMAX(1)=1.23, CMO=0.0151, LERI=.008483, CLAMO=0.0151$
 $WGSCHR CLMAXL=1.23$
 DAMP
 CASEID INCLUDES BODY AND WING-BODY EXPERIMENTAL DATA, EX.PROB. 3, CASE 3
 SAVE
 NEXT CASE
 $FLTCON NMACH=1.0, MACH(1)=0.1, RNNUB(1)=5.622E4$
  $PROPWR AIETLP=0.0,NENGSP=2.0,THSTCP=1.1893,
  PHALOC=0.0, PHVLOC=0.0, PRPRAD=1.75,
  ENGFCT=60.0, NOPBPE=2.0,BAPR75=15.0, YP=13.195, CROT=.TRUE.$
 DAMP
 CASEID INCLUDES BODY AND WING-BODY EXPERIMENTAL DATA, EX.PROB. 3, CASE 4
 NEXT CASE
1       THE FOLLOWING IS A LIST OF ALL INPUT CARDS FOR THIS CASE.
0
 BUILD
 $FLTCON  NMACH=1.0, MACH(1)=0.1,
  NALPHA=9.0, ALSCHD(1)=-2.0,0.0,2.0,4.0,8.0,12.0,16.0,20.0,24.0,
  RNNUB(1)=5.622E4$
 $OPTINS SREF=528.0, CBARR=6.139, BLREF=86.0$
 $SYNTHS XCG=1.16641, ZCG=0.0, XW=0.0, ZW=0.0, ALIW=0.0, XH=0.0,
  ZH=0.0, ALIH=0.0, XV=0.0, VERTUP=.TRUE.$
 $WGPLNF CHRDTP=6.139, SSPNE=43.0, SSPN=43.0, CHRDR=6.139,
  SAVSI=0.0, CHSTAT=0.0, SWAFP=0.0, TWISTA=0.0, SSPNDD=33.3,
  DHDADI=2.0,DHDADO=3.0, TYPE=1.0$
 $WGSCHR TOVC=0.0931, DELTAY=0.00497, XOVC=0.0931, CLI=0.60, ALPHAI=4.80,
  CLALPA(1)=0.1075, CLMAX(1)=1.23, CMO=0.0151, LERI=.008483, CLAMO=0.0151$
  $WGSCHR CLMAXL=1.23$
 DAMP
 CASEID INCLUDES BODY AND WING-BODY EXPERIMENTAL DATA, EX.PROB. 3, CASE 3
 SAVE
 NEXT CASE
0 INPUT DIMENSIONS ARE IN FT, SCALE FACTOR IS 1.0000
```

AUTOMATED STABILITY AND CONTROL METHODS PER APRIL 1976 VERSION OF DATCOM
CHARACTERISTICS AT ANGLE OF ATTACK AND IN SIDESLIP
WING ALONE CONFIGURATION
INCLUDES BODY AND WING-BODY EXPERIMENTAL DATA, EX.PROB. 3, CASE 3

---------------------- FLIGHT CONDITIONS ------------------------ -------------- REFERENCE DIMENSIONS ------------

MACH NUMBER	ALTITUDE	VELOCITY	PRESSURE	TEMPERATURE	REYNOLDS NUMBER	REF. AREA	REFERENCE LONG.	LAT.	REF. LENGTH HORIZ	MOMENT REF. CENTER VERT
	FT	FT/SEC	LB/FT**2	DEG R	1/FT	FT**2	FT	FT	FT	FT
0 0.100					5.6220E+04	528.000	6.139	86.000	1.166	0.000

0   -------------------DERIVATIVE (PER DEGREE)-------------------

ALPHA	CD	CL	CM	CN	CA	XCP	CLA	CMA	CYB	CNB	CLB
-2.0	0.014	-0.114	0.0201	-0.115	0.010	-0.175	9.349E-02	-5.610E-03	-2.785E-04	1.291E-06	-1.022E-03
0.0	0.014	0.073	0.0088	0.073	0.014	0.120	9.425E-02	-5.659E-03	-2.785E-04	5.298E-07	-1.204E-03
2.0	0.016	0.263	-0.0026	0.263	0.006	-0.010	9.493E-02	-5.712E-03	-2.785E-04	6.800E-06	-1.388E-03
4.0	0.019	0.453	-0.0140	0.453	-0.013	-0.031	9.442E-02	-5.719E-03	-2.785E-04	2.024E-05	-1.573E-03
8.0	0.031	0.824	-0.0367	0.821	-0.084	-0.045	7.814E-02	-4.860E-03	-2.785E-04	6.702E-05	-1.933E-03
12.0	0.043	1.078	-0.0529	1.063	-0.182	-0.050	2.128E-02	-1.517E-03	-2.785E-04	1.146E-04	-2.179E-03
16.0	0.039	0.995	NA	0.967	-0.237	NA	-3.200E-02	NA	-2.795E-04	9.756E-05	-2.098E-03
20.0	0.031	0.822	NA	0.783	-0.252	NA	-2.738E-02	NA	-2.813E-04	6.666E-05	-1.931E-03
24.0	0.029	0.776	NA	0.720	-0.289	NA	4.112E-03	NA	-2.825E-04	5.933E-05	-1.885E-03

AUTOMATED STABILITY AND CONTROL METHODS PER APRIL 1976 VERSION OF DATCOM
DYNAMIC DERIVATIVES
WING ALONE CONFIGURATION
NCLUDES BODY AND WING-BODY EXPERIMENTAL DATA, EX.PROB. 3, CASE 3

---------------------- FLIGHT CONDITIONS ------------------------ -------------- REFERENCE DIMENSIONS ------------

MACH NUMBER	ALTITUDE	VELOCITY	PRESSURE	TEMPERATURE	REYNOLDS NUMBER	REF. AREA	REFERENCE LONG.	LAT.	REF. LENGTH HORIZ	MOMENT REF. CENTER VERT
	FT	FT/SEC	LB/FT**2	DEG R	1/FT	FT**2	FT	FT	FT	FT
0 0.100					5.6220E+04	528.000	6.139	86.000	1.166	0.000

DYNAMIC DERIVATIVES (PER DEGREE)

0   -------PITCHING------- -----ACCELERATION------ --------------ROLLING-------------- --------YAWING--------

ALPHA	CLQ	CMQ	CLAD	CMAD	CLP	CYP	CNP	CNR	CLR
-2.00	5.797E-02	-1.180E-02	NDM	NDM	-9.819E-03	-1.566E-03	2.516E-04	-7.401E-05	-5.742E-04

0.00	-9.900E-03	-1.341E-03	-1.611E-04	-7.205E-05
3.678E-04				
2.00	-9.979E-03	-1.114E-03	-5.772E-04	-8.816E-05
1.318E-03				
4.00	-9.942E-03	-8.854E-04	-9.962E-04	-1.227E-04
2.273E-03				
8.00	-8.294E-03	-4.381E-04	-1.823E-03	-2.429E-04
4.137E-03				
12.00	-2.400E-03	-1.226E-04	-2.423E-03	-3.652E-04
5.410E-03				
16.00	3.200E-03	-1.999E-04	-2.267E-03	-3.214E-04
4.991E-03				
20.00	2.755E-03	-4.140E-04	-1.870E-03	-2.420E-04
4.126E-03				
24.00	-5.338E-04	-4.811E-04	-1.761E-03	-2.231E-04
3.892E-03				

0*** NDM PRINTED WHEN NO DATCOM METHODS EXIST
1   THE FOLLOWING IS A LIST OF ALL INPUT CARDS FOR THIS CASE.
 $FLTCON NMACH=1.0, MACH(1)=0.1, RNNUB(1)=5.622E4$
 $PROPWR AIETLP=0.0,NENGSP=2.0,THSTCP=1.1893,
 PHALOC=0.0, PHVLOC=0.0, PRPRAD=1.75,
 ENGFCT=60.0, NOPBPE=2.0,BAPR75=15.0, YP=13.195, CROT=.TRUE.$
DAMP
CASEID INCLUDES BODY AND WING-BODY EXPERIMENTAL DATA, EX.PROB. 3, CASE 4
NEXT CASE
0 INPUT DIMENSIONS ARE IN FT, SCALE FACTOR IS 1.0000

AUTOMATED STABILITY AND CONTROL METHODS PER APRIL 1976 VERSION OF DATCOM
CHARACTERISTICS AT ANGLE OF ATTACK AND IN SIDESLIP
WING ALONE CONFIGURATION
INCLUDES BODY AND WING-BODY EXPERIMENTAL DATA, EX.PROB. 3, CASE 4

---------------------- FLIGHT CONDITIONS ----------------------   -------------- REFERENCE
DIMENSIONS ------------
 MACH   ALTITUDE VELOCITY   PRESSURE   TEMPERATURE   REYNOLDS   REF.   REFERENCE
LENGTH   MOMENT REF. CENTER
 NUMBER   NUMBER   AREA   LONG.   LAT.   HORIZ
VERT
         FT   FT/SEC   LB/FT**2   DEG R   1/FT   FT**2   FT   FT   FT   FT
0 0.100   5.6220E+04   528.000   6.139   86.000   1.166
0.000
0   ------------------DERIVATIVE (PER DEGREE)------------------
0 ALPHA   CD   CL   CM   CN   CA   XCP   CLA   CMA   CYB   CNB
CLB
0
 -2.0 0.014 -0.114   0.0201 -0.115   0.010 -0.175   9.349E-02 -5.610E-03 -2.785E-04   1.291E-06 -
1.022E-03
  0.0 0.014   0.073   0.0088   0.073   0.014   0.120   9.425E-02 -5.659E-03 -2.785E-04   5.298E-07 -1.204E-
03
  2.0 0.016   0.263 -0.0026   0.263   0.006 -0.010   9.493E-02 -5.712E-03 -2.785E-04   6.800E-06 -
1.388E-03
  4.0 0.019   0.453 -0.0140   0.453 -0.013 -0.031   9.442E-02 -5.719E-03 -2.785E-04   2.024E-05 -
1.573E-03
  8.0 0.031   0.824 -0.0367   0.821 -0.084 -0.045   7.814E-02 -4.860E-03 -2.785E-04   6.702E-05 -
1.933E-03
 12.0 0.043   1.078 -0.0529   1.063 -0.182 -0.050   2.128E-02 -1.517E-03 -2.785E-04   1.146E-04 -
2.179E-03

```
   16.0  0.039  0.995   NA   0.967  -0.237   NA    -3.200E-02   NA    -2.795E-04  9.756E-05  -2.098E-
03
   20.0  0.031  0.822   NA   0.783  -0.252   NA    -2.738E-02   NA    -2.813E-04  6.666E-05  -1.931E-
03
   24.0  0.029  0.776   NA   0.720  -0.289   NA     4.112E-03   NA    -2.825E-04  5.933E-05  -1.885E-
03
```

0*** NA PRINTED WHEN METHOD NOT APPLICABLE

AUTOMATED STABILITY AND CONTROL METHODS PER APRIL 1976 VERSION OF DATCOM
DYNAMIC DERIVATIVES
WING ALONE CONFIGURATION
INCLUDES BODY AND WING-BODY EXPERIMENTAL DATA, EX.PROB. 3, CASE 4

```
----------------------- FLIGHT CONDITIONS -----------------------      ------------- REFERENCE DIMENSIONS -------
-----
  MACH  ALTITUDE VELOCITY  PRESSURE  TEMPERATURE   REYNOLDS        REF.    REFERENCE LENGTH
MOMENT REF. CENTER
 NUMBER                                 NUMBER       AREA    LONG.   LAT.   HORIZ   VERT
         FT      FT/SEC   LB/FT**2   DEG R    1/FT      FT**2    FT      FT     FT      FT
0 0.100                                      5.6220E+04      528.000   6.139   86.000  1.166   0.000
                                       DYNAMIC DERIVATIVES (PER DEGREE)
0    ------PITCHING------  ----ACCELERATION------  --------------ROLLING-------------  --------YAWING--------
0 ALPHA    CLQ      CMQ      CLAD      CMAD      CLP       CYP       CNP       CNR       CLR
0
  -2.00  5.797E-02 -1.180E-02   NDM       NDM    -9.819E-03 -1.566E-03  2.516E-04 -7.401E-05  -
5.742E-04
   0.00                             -9.900E-03 -1.341E-03 -1.611E-04 -7.205E-05  3.678E-04
   2.00                             -9.979E-03 -1.114E-03 -5.772E-04 -8.816E-05  1.318E-03
   4.00                             -9.942E-03 -8.854E-04 -9.962E-04 -1.227E-04  2.273E-03
   8.00                             -8.294E-03 -4.381E-04 -1.823E-03 -2.429E-04  4.137E-03
  12.00                             -2.400E-03 -1.226E-04 -2.423E-03 -3.652E-04  5.410E-03
  16.00                              3.200E-03 -1.999E-04 -2.267E-03 -3.214E-04  4.991E-03
  20.00                              2.755E-03 -4.140E-04 -1.870E-03 -2.420E-04  4.126E-03
  24.00                             -5.338E-04 -4.811E-04 -1.761E-03 -2.231E-04  3.892E-03
0*** NDM PRINTED WHEN NO DATCOM METHODS EXIST
1      THE FOLLOWING IS A LIST OF ALL INPUT CARDS FOR THIS CASE.
0
1 END OF JOB.
```

22.0 References

Research Papers

- *"Development of an efficient solar powered UAV with onboard solar tracker"* By Troy Tegeder
- *"SKY-SAILOR Design of an autonomous plane"* By ANE A. Noth, S. Bouabdallah, S. Michaud, R. Siegwart, W. Engel
- *"Design of Solar Powered Airplanes for Continuous Flight"* By A. Noth, R. Siegwart, W. Engel
- *"Design and Analysis of Low Reynolds Number Airfoils"* By Nicholas K. Borer
- *"NASA's High Altitude Long Endurance UAV Analysis of Alternatives and Technology Requirements Development"* By Craig L. Nickol and Mark D. Guynn, Lisa L. Kohout, Thomas A. Ozoroski
- *"Design of an Unmanned Aerial Vehicle by Conceptual Approach"* By Fahad Aman Khan Amina Malik
- *"EP-UAV Project 2006 , The Kingfisher" joint project* By RMIT and NUAA Universities.
- *"History of Solar flight"* By André Noth
- *"Analysis of wing slipstream flow interaction"* By Antony Jameson
- *"Reduction in efficiency of propellers due to slipstream"* By Max M Munk
- *"HELIPAT HALE design project" By* Politecnico di Torino University, Itlay
- *"Structural and Manufacturing Analysis for Meridian UAV Wing Concept"* By University of Kansas, Technical Report CReSIS TR 121
- *"Notes on Aerospace Structures"* By Edmundo Corona
- *"Aerodynamic Design Studies of Conventional & Unconventional Wings with Winglets "* By Dr. R. K. Nangia Dr. M. E. Palmer Mr. R. H. Doe

Books

- Daniel P. Raymer. (1999). Aircraft Design: *"A Conceptual Approach"* – 3rd Edition. AIAA Educational Series

- Jan Roskam. (2000). *Airplane Design: Part I-VII*. AIAA Educational Series

- Jane's All the World UAV and Target Drones. (2002). Janes Defence Publications

- J. D. Anderson. (2004). *Introduction to Flight.* AIAA Educational Series

- J. D. Anderson. (2005). *Aircraft Performance.* McGraw Hill Publications

- Daniel P. Raymer. (2003). *Simplified Aircraft Design for Homebuilders.* Design Dimension Press

- Martin E. Eshelby. (2000). *Aircraft Performance Theory and Practice.* Butterworth-Heinemann

Websites

- www.vectorsite.net/twuav_12.html
- www.globalsolar.com
- www.mysolarstop.com/cart/flexible-solar-panels-c-1.html
- www.siliconsolar.com/visual-directory/flexible-solar-panels.html
- www.en.wikipedia.org/wiki/NASA_Pathfinder
- www.nasa.gov/centers/dryden/news/FactSheets/FS-034-DFRC.html
- www.aerospacemetals.com/aluminum.html#rod
- www.blazingwings.org/entry/top-12-solar-powered-aircrafts/
- www.en.wikipedia.org/wiki/Electric_motors
- www.hackerbrushless.com
- www.me.ua.edu/me465/PDF/Motor_Selection_Sizing1.pdf
- www.paso.esa.int/5_training_materials/training_12_aerospace%20applic.pdf
- www.4-max.co.uk/pdf/Electric%20Flight%20Glossary.pdf
- www.ingenia-cat.com/reference/learn/TEC.PAP.7055681008.pdf

Softwares

Advance Aircraft Analysis (AAA), MatLab 2007, Design Foil, Profili2, Datcom, ANSYS 11, Gridgen, FLUENT, Aerodynamics Software, Rc-CAD, AutoCAD, Office 2007, RDS